The Assassin Indianapolis

JIM WEST

Other Books by Jim West

DNAlien

DNAlien II

DNAlien III

Genocide by GMO

Living Within a Strange Mind Vol I

Living Within a Strange Mind Vol II

The Making of an Assassin Atlanta

The Assassin Baltimore

The Assassin Chicago

The Assassin Denver

The Assassin El Paso

The Assassin Fort Worth

The Assassin Galveston

The Assassin Honolulu

As always, thanks to my friend John Fleenor, whose sarcastic comments and ever-ready wit keep me smiling while I make the necessary corrections to my terrible grammar, punctuation, or sentence structure.

Also, for pointing out obvious errors when I marry two complete strangers or kill an unknown character for no apparent reason.

John has been the common thread throughout all of my books and tried in vain to get me to produce at least one enjoyable book that is also factual and coherent.

As long as he suffers through my efforts, I'll keep trying to entertain my minuscule but welcome reading audience.

Thank you, John!!!

Chapter 1

Jim Lashley had barely entered his house in Mesquite, Texas, after completing a three-day trip with American Airlines when the phone rang.

"Hello," Jim said, setting his suitcase down beside the end table where the phone was located in the living room.

"Good afternoon, Jim," the caller said. "How was the trip to wherever you were for the last three days?"

"Pretty much standard," Jim answered, recognizing the voice of General Gene Barker. "Leave Dallas with 160 close friends, drop them off at some town, pick up another 150 or so close friends and take them somewhere, swap them for a different group of 155 close friends and take them somewhere.

"Then spend the night in some luxurious accommodations where the company has cut-rate rooms for its crews," Jim continued. "Then get up early, repeat the routine a couple of more times, and another hotel. Finally, bring a different group of close friends back to Dallas and come home for a too-short few days, and then back to transporting my closest friends around the United States for three more days."

"You chose that path, my boy," Gene said, chuckling. "But I bet it still beats pushing those smelly F4s around the skies for that famous uncle of yours. Not to mention having a toilet located just outside of the cockpit door and a beautiful young lady catering to your every desire."

"I'll agree about the toilet, but I don't think you've flown commercial in many years," Jim replied. Otherwise, you'd know that those beautiful young ladies, from 40 years ago, aren't the same as you remember. And let's not get into the pretty young boys that seem to have encroached themselves on the flight schedules.

"But I'm hazarding a guess that you aren't calling just to discuss my endeavor to pay the rent and put food on the table," Jim finished. "So, to what do I owe the pleasure of this call?"

"Well, if you'd go open the door, I'll discuss it with you," Gene said, ringing the doorbell and hanging up.

Jim replaced the phone, walked the few steps to the door, and opened it to see Gene standing there smiling. "Please, come in," Jim told him, turning back into the living room. "Would you possibly be interested in a cold beer, or something stronger?"

"Something stronger," Gene answered, spotting the suitcase by the end table. "I'm betting three drinks that you barely beat me to the house."

"I don't think I'll take that bet," Jim told him, leading the way into the kitchen. "I'll bet you had me tracked by that ill-tempered computer hacker Bracer and knew exactly where I was every foot of my way from the airport, but I'll gladly join you in the three drinks, possibly four if we don't have anything philosophical to discuss."

"Just two old friends catching up on our lives," Gene answered, following him. "By the way, how is Marie?"

"She's just fine," Jim said, reaching into the top shelf of a cabinet and getting a bottle of Jack Daniel's. "She'll be sorry she missed you. That is, if you're in a hurry to leave after the drinks and *catching up*."

"I can spare an hour or two," Gene answered, taking a bottle of Coke from the refrigerator. "Possibly enough time to take both of you out to dinner. That is if her dad is still making his famous lasagna."

"I can guarantee he is," Jim said, getting two glasses down and filling them with ice. "But if we give him a call, I'm sure he'll whip up something special for us."

"Not necessary," Gene said, following Jim out to the back porch. "I don't ever want to be an imposition on the poor man. After all, he has you mooching free meals and chasing his daughter. That burden must weigh tremendously on him every day. Let me bring a little ray of sunshine into his life. And heighten it with the happy surprise of seeing me."

"The high point of his dreary life, I'm sure," Jim said, nodding to one of the wrought iron chairs and setting the bottle and glasses on the table.

Jim waited until Gene took his seat and sat across the table and then poured each glass half full of Jack Daniel's and passed one to Gene.

As Gene added Coke to his and handed Jim the bottle, he waited until Jim had filled his glass.

Pouring the Coke into his glass, Jim then raised it and said, "Semper fi."

"Semper fi," Gene echoed and tapped his glass with Jim's.

Taking a short sip, Jim said, "If you'll pardon me for a couple of minutes, I'd like to get out of this monkey suit the airlines call a uniform and put on something a little more comfortable."

"Take your time," Gene said, nodding. "I'm pretty damn happy just sitting here enjoying the weather and the drink you so graciously provided."

A few minutes later, Jim came back out wearing his customary Wrangler jeans and a T-shirt. Picking up his glass after taking his seat, he asked, "Okay. Now that the pleasantries are done to satisfy the rules of decorum, just what really brings you down here?"

Gene sat quietly, looking at Jim for a moment, and then answered, "What do you know about the Chinese Triad?"

"Not much," Jim admitted. "Some sort of Chinese gang or cartel, I think."

"More or less," Gene agreed. "But the more is that they date back to secret societies of the Qing dynasty in China. They are a family-run organized crime gang and operate pretty much globally. More on that later. Now, what do you know about a group known as Tren de Aragua?"

"Pretty much the same," Jim answered. "Some gang down in Argentina, I think."

"Venezuela," Gene corrected him. "But no longer restricted to Venezuela. They are now a transnational criminal organization. And our interest is in their activities here in the United States."

"Let me add two and two," Jim replied, taking another sip of his drink. "You, or Black Water, have something to do with both gangs. And you're here to either ask my advice, which is obviously nil, or to tell me that I'm being asked to become involved."

"Very perceptive, Grasshopper," Gene said, nodding. "But then you've always been able to discern the intuitive obvious. Before I continue, perhaps a refill?"

Chapter 2

When Jim returned with fresh ice in the glasses, Gene asked, "How informed are you regarding the illegal immigration issue?"

"No more than the average, I would say," Jim answered as he refilled his glass. "I'm going out on a limb and say that both the Triad and this Aragua groups are part of the immigration issue."

"Once again, your skills at connecting the dots are impressive," Gene remarked, taking the bottle of Jack Daniel's from Jim. "To more or less start over, we, meaning Black Water, have been given a task much like we had down in El Paso a few years back.

"Now, don't get me wrong," Gene continued adding Coke to his drink, "As close to the MS 13 problem as this is, it's complicated by the fact that most people understand the threat Tren de Aragua presents to their communities, but many don't exactly concur with the idea of mass deportation that the incoming President has espoused.

"That, and the political ramifications of potentially several hundred or possibly even thousands of deaths, have to be worked as covertly as possible to prevent the uproar from the outgoing political party," he explained. "That is the primary reason that the incoming Border Czar has turned this part of the problem over to us.

"And one of the issues that we will be dealing with is with the children that the Triad has smuggled into the country," Gene concluded.

"Let me catch up here," Jim said, sitting back in his chair. "You're not saying that there will be hundreds of deaths of these children, are you?"

"No," Gene answered. "It's possible that some may be injured during the operation, but our goal, especially with the Chinese children, is to send them back to their country of origin. Tren de Aragua is a different matter.

"Much like MS 13, there are some very ruthless actors in that group," Gene explained. "I'm sure you remember the ages of some of the MS 13 gangs. Same here with Tren de Aragua. There are some that don't even need to shave yet, but they have no qualms about shooting anyone who opposes them.

"The best thing I can tell you is that there aren't that many under the age of 15," Gene told him. "And we'll attempt to take them to a collection point and turn them over to the Homeland Security Department. We're working with less than one hundred Tren de Aragua members in your area and about the same with the other operations across the country.

"You say *my area* as if the decision has already been made," Jim remarked, looking at Gene. "Nor have you explained where *my area* happens to be, when I'm supposed to begin operations in *my area*, or anything other than that the country has an immigration problem, especially concerning Venezuela and China."

"Forgive me," Gene replied. "I'll rephrase. I'm asking you to look at this as you have each operation Black Water has presented to you. It's always your option to refuse any operation, just as it's been since you joined the company.

"And I was definitely going to tell you about your area, should you accept the assignment," Gene continued. "But I got a little ahead of myself. Your area, if you accept, will be Indianapolis.

"You know we would never force you to take an operation you didn't feel comfortable with," Gene reminded him. "That's always been the company's position. But you also know that we would never accept a contract unless we were sure it was absolutely necessary."

"Of course," Jim told him contritely. "I was a little preemptive in my remark. I trust Black Water. And that is because I trust you. You've always given me opportunities that have been instrumental in my life.

"From the jungle floors of Vietnam, to earning my wings as a Marine aviator, and now flying for American, you've been behind me every step," Jim continued. "And I appreciate your support more than you'll ever know. So, if you'll forgive my outburst, I'll pour us another small drink before we go to dinner."

"Dinner?" Gene demanded, mocking Jim. "When did I agree to go to dinner? And just exactly where are you taking me?"

"Touche," Jim said, smiling as he raised his glass. "Except that we did discuss you taking Marie and me to dinner. And, if you'll reach back into the recesses of your mind, you'll see that you alluded to her father making lasagna."

"You are correct," Gene replied, tapping Jim's glass with his. "Now, if you'll bear with me for another few minutes, I'll try to wrap up this overview, and we can head for the restaurant."

Seeing Jim's nod, he continued, "Tren de Aragua, or TDA, started in a prison in Venezuela, and has spread across our hemisphere. They more or less specialize in human trafficking and drug smuggling. That's not to say that there isn't plenty of petty theft, street robberies, prostitution, and shootings.

"However, they aren't quite as established as MS 13," Gene continued. "Mainly because they haven't been in our country as long. But they are growing. Rapidly. Especially with the millions of illegals that have flooded our country in the last four years. They have been particularly active in recruiting new members from the migrant shelters.

"The other half of the operation involves the Chinese issue," he explained. "They are more widely dispersed and entrenched in our country. Again, because of the element of time.

"They have become so well organized that they have their own corporation here in the US," he continued. "Their corporation, named 'Triad Agriculture and Land Development', is buying farmland, strip malls, and other operations across the country.

"Most of the time, the farmland is located in the vicinity of a military facility," Gene said. "And most of the time, it is also a marijuana operation. Even though marijuana is legal in most of the states, the production is restricted to varying degrees depending upon the state.

"However, none of the facilities operated by the Triad are legal," he concluded. "And that is part of where the trafficking of children plays into the problem.

"The other part of the child smuggling operation involves the strip malls owned and operated by the Triad. More accurately, certain operations within the strip malls.

"The boys smuggled in usually end up working on the farms," Gene explained. "But the girls end up working in the massage parlors initially. And we're talking about some as young as twelve.

"Once they've served their useful purpose there, they are transferred to nail or beauty salons, and then to hotel maid service," Gene continued. "Now, we don't expect much violence with that portion of the operation. Our job will be to gather the children up and deliver them to the Homeland Security or Immigration and Customs Enforcement, ICE."

"That sounds like this could take some time," Jim remarked, setting his empty glass on the table. "I can't imagine how long it's going to take just finding these people and developing a plan to implement the operation."

"Most of that has already been done," Gene told him, placing his glass on the table and standing. "We were contacted almost a year ago and asked to provide the necessary support to accomplish the incoming President's promise to resolve the illegal immigration problem."

"How could you have contracted with the new administration when the election was only decided a couple of months ago?" Jim asked as they carried their glasses back into the kitchen.

"Let's just say that some very powerful political people made arrangements for Black Water to submit a concept of operations," Gene answered, putting his glass in the sink.

"What would have happened if the election had gone the other way?" Jim asked as they headed for the door.

"The company would have spent millions of dollars and be left paying the tab," Gene answered. "Let's never dwell on what could have been … concentrate on what is. I'm sure you remember my saying regarding the age-old question, 'Why did the chicken cross the road?' It doesn't matter why; we're dealing with a chicken who is on the other side of the road and that's what's important."

"Pardon me for correcting you, sir," Jim said as they headed for Gene's typical choice of a black Suburban. "But I believe I told you that and credited my great-great-grandfather with the saying."

"You could be right," Gene replied, smiling at Jim. "How about we take your little 'Vette instead of my Suburban?"

Jim stopped and looked at Gene, saying, "I can't believe it. You're opting for a ride in what you usually refer to as the most uncomfortable piece of shit you've ever been in."

"Let's just call it a small concession for referring to *your area* without asking if you were interested," Gene told him as Jim opened the garage door. "And, if you'll remember, it was over a hundred degrees the time I rode with you and there's no air conditioner in that car."

"One final question before we get to Siciliano's," Jim said as they got into the car. "When do we start?"

Chapter 3

"We'll start on January 20, as soon as the new President is sworn in," Gene told him as they headed for Garland.

"Don't we have to wait until the Senate approves his cabinet?" Jim asked as they merged onto 635 heading north.

"The new *Border Czar* doesn't need Senate approval since he doesn't hold a cabinet position," Gene answered, watching Jim swerve in and out in the increasing traffic. "He is ready to hit the ground running once the President gives him the nod."

"All right," Jim said a few moments later as they took the exit for Shiloh Road, northbound. "What about all the groundwork to identify our targets? Not to mention all of the vehicles, briefing for the teams, and whatever innovative technology Bracer has managed to devise?"

"Most of that is in place," Gene assured him as they turned right onto Buckingham Road. "The only thing we need to do for now is get all of the fifty team leaders together at Quantico and cover both the new toys Bracer has ready and discuss any issues that may have arisen since I left there a week ago."

"What have you been doing for the last week before you came to see me?" Jim asked as they spotted Siciliano's A Taste of Italy ahead on the right.

"Meeting with the other team leaders regarding their *areas,*" Gene answered. "Oh, by the way. I need you to come with me tomorrow and join us."

"I just get home after being gone for three days and you want me to up and leave?" Jim asked, astonished. "Don't you think I'd like a day or so with Marie before I run off?"

"I've got that covered," Gene told him as they parked. "I believe that if you approached the issue correctly, Marie won't have too many objections."

"And what's your idea of the correct way to tell her that I can't be home with her for the few days before American sends me back out for another three days?" Jim asked as they headed for the door.

"Invite her to come with us," Gene said, smiling as he opened the door for Jim. "Little vacation for her, and she gets to spend your precious days off with her."

Jim just shook his head as they walked in. Almost immediately, Marie spotted them and rushed over, throwing her arms around Jim's neck, saying, "I was just about to call you! I've been waiting to hear from you when you were supposed to be home."

"Slight delay," Jim told her as he kissed her quickly.

Marie released Jim and stepped over to Gene, saying, "I can guess what caused the slight delay."

Giving Gene a hug, she continued, "And I'd bet that delay included a sip or two of Jack Daniel's that I got a whiff of."

"Good to see you, Marie," Gene said as Marie stepped back. "Beautiful as ever. And how are those two sweet daughters?"

"They're just fine," she answered. "Back at school and enjoying the college life. Now, I'm going out on a limb here and getting you a table away from the otherwise normal people who enjoy dining here."

"That would be great," Jim said, following her to the rear of the restaurant. "And, if it's okay with Tony, I'd love to have you join us."

"I'll see if I can convince Dad to let me have my union contract hour dinner break," she said, smiling as they took their seats. "I'll be right back with some bread and marinara sauce. Two unsweet iced teas with lemon?"

"As always," Jim replied, pushing his menu away. "And I believe we are ready to order any time that's convenient for you."

"Let me guess," Marie answered, taking all the menus from the table. "Should I tell Dad two lasagnas?"

"Works for both of us," Jim said, smiling as Marie put her hand on his shoulder. "And please, ask Tony to stop by when he has a chance."

As Marie headed back to the kitchen, Jim looked at Gene and said, "I think maybe you should be the one to inform her about your vacation plan for us."

"You think I should be the one?" Gene asked, grinning. "What's that you always say? Oh yeah. 'If that's what you think, maybe you need a little more practice thinking so you don't keep making a fool of yourself.'"

"Close enough," Jim said, nodding. "But it was my great-great-grandfather who told me that every time I was wrong about what I thought."

"Your great-great-grandfather sure was an amazing philosopher," Gene retorted. "Seems he had a thoughtful saying for just about everything."

"That he did, that he did. And now I'm the most highly respected and wisest sage of modern times, thanks to his wisdom," Jim said as Marie came back with a tray and Tony close behind.

"My friends!" Tony exclaimed as Marie set the tray on the table and both Jim and Gene stood to greet him.

Giving Jim a hug, Tony said, "How was your trip?"

"Boring as usual," Jim said, returning the hug. "So glad to be home."

"And you, Gene," Tony said, stepping over to him. "It's been too long since you visited us."

"Work, work, work," Gene said as Tony gave him a quick hug. "But I'm here today for some of your famous lasagna. And hopefully a little of your pleasant company, of course."

"I hope you'll allow your beautiful daughter to join us for dinner," Jim said as Tony motioned for them to have a seat. "And it would be our pleasure if both you and your amazing wife, Aurora, could join us for a drink later."

"I believe I can arrange that," Tony said as Aurora entered the room. "I happen to know the management quite well."

"What's this *amazing wife* comment?" Aurora said, putting her arms around Jim. "Marie is beautiful and I'm simply amazing?"

"I make a conscious effort to never let a husband know that I find his wife beautiful," Jim said as Aurora stepped back. "You just never know when a streak of jealousy might arise."

"You may consider my lovely wife beautiful anytime you wish," Tony said, putting his arm across her shoulders. "And I shall always consider her beautiful."

"Fair enough," Jim said, putting his arm around Marie, "we both have beautiful ladies, and they shall remain beautiful forever."

"Now, I must return to that hell I call a kitchen and prepare the finest lasagna known around the world for my friends," Tony said before turning. "And my daughter Marie will provide you with a bottle of my special reserve Merlot to enjoy with your meals. And I shall join you with my amazing wife shortly after you finish."

Chapter 4

After they had left, Jim took a slice of the warm bread, dipped it in the marinara sauce, and asked, "What do we do with our targets after we find them? More specifically, what do we do with the bodies of the unfortunate few who refused our offer of extradition?"

"They will be taken to the same place where Homeland Security, or ICE, will take responsibility," Gene answered. "That then becomes their problem. I do know that a certain portion of Guantanamo is being prepared, mainly for the TDA folks, but probably any of the illegals that are suspected of breaking any of our laws but not arrested, other than illegal entry."

"I thought Guantanamo, or GITMO, had been closed due to repeated reports of 'inhumane treatment,'" Jim said, watching Gene take a slice of bread and smear butter across the top.

"Previous administrations made some attempts," Gene answered. "But wiser men in Congress saw the need and kept the facility operational, although with a much smaller population.

"GITMO is still the best place for these TDA characters," Gene finished. "I damn sure don't want them next door to my family, or any family for that matter. GITMO was set up for this sort of detention, and I can't think of any reason not to send them there."

"I was under the impression that all of the illegals were to be deported to their country of origin," Jim said. "Doesn't this constitute some issue with the right to a trial?"

"Actually, no," Gene answered. "It can be argued that an illegal immigrant doesn't have rights under our constitution unless they have violated the law in the US. Any of them who have been charged with criminality, tried, and found guilty will be held in our prison system.

"The ones we're deporting are just being sent there awaiting transportation. They aren't citizens and technically don't have the same rights as one. And since he, or she, is not a combatant by normal standards, that is wearing an identifiable uniform, they don't fall under the Geneva Convention regarding treatment of prisoners of war, or POWs," Gene explained.

"This has been argued back and forth for years, and several lawsuits have been filed on behalf of GITMO prisoners," Gene finished. "Regardless, it isn't our problem. We just gather up the trash and leave it at their curb, ICE or Homeland Security's curb."

"One other thing that has been on my mind," Jim started as he looked around to ensure there was no one close enough to hear. "Why can't ICE and the other agencies take care of this?"

"Several reasons that are way above your pay grade, and mine," Gene answered. "First, lack of personnel. At least a lack of sufficient personnel to remove over ten million illegals within the time frame the administration wants.

"You've got to remember that this is both a national security issue and a political one," Gene continued. "And as most hot-button political issues, this one can't drag on and on. So, the decision was made somewhere in the higher echelons of power that the incoming President wanted this accomplished within the first sixty days of his administration.

"And before you say anything, the emphasis on the order for now is those who are currently involved with illegal activities, such as the human trafficking and child sex slave trade that the Triad is conducting, or other heinous crimes conducted by TDA," Gene said. "Those who are only guilty of illegal entry will take longer, but the emphasis, and our contract, does not involve this group.

"Out of the ten or so million illegals, most, let's say ninety percent, aren't on the initial deportation list," Gene clarified. "I'm not sure, but I don't really expect Black Water to be involved in the 'non-violent illegal' problem."

"Can GITMO hold a million people?" Jim asked. "If you're talking about ten percent of ten million, that's one million. I didn't think GITMO ever held more than a thousand or so."

"That's true," Gene admitted. "There are several 'camps' throughout the facility that hold up to about a thousand. However, there is one known as *Camp No* that is located outside the camp perimeter and I'm not sure how large it can be expanded.

"Having said that, you also have to remember that most of these folks are going to be loaded on a plane and shipped back to Colombia, or Haiti, or Venezuela, as quickly as possible," Gene concluded. "Not to mention the majority who aren't guilty of more than theft, or other non-violent crime, will probably be flown out of one of the many airports within the US."

"I guess that would mean a large charter operation, such as Air America back during Vietnam," Jim suggested. "I'm sure the normal flight from DFW to Sao Paulo wouldn't allow the folks we're discussing."

"Probably not," Gene agreed. "But that's not your problem. Nor mine. I'd imagine that the CIA has already been busy lining up coach-only accommodations for the unwanted noncitizens currently residing in our country."

"You know what bothers me?" Jim asked. "It's the people who may have entered the country legally and have, for some reason or other, overstayed their visa. This is going to be such a massive operation with such a short fuse; some of them may well be caught up in the broad net we're casting. And we both know that our immigration system can't handle the normal shit. This has the potential to become a rubber stamp for any person whose name crosses some low-level bureaucratic fat-assed schmo's desk who doesn't give a rat's ass about anything other than being out of his office at the stroke of five."

"I wish I could disagree with you," Gene said as he saw Marie coming with a bottle of wine and Aurora right behind her carrying a tray of steaming lasagna. "But again, above my pay grade."

"Ready for the best lasagna in the Metroplex?" Aurora said, setting the tray in the middle of the table.

"Indeed," Jim said as Marie poured a glass of wine and handed it to him.

"I've been waiting months for this," Gene said, smiling. "I don't get to come here as often as I wish.'

"You can't blame us for that," Marie said, handing him a glass. "If I'm not mistaken, you have access to a private jet that can fly down here any time you wish."

"Oh, that would work for maybe a couple of months," Gene told her as she sat down. "Then I'd be working as a Walmart greeter somewhere in lower Alabama and you'd never get to see me again."

"I'd make Tony hire you as Siciliano's official greeter," Aurora said as she looked around the table. "Then you'd become almost family, like our dear friend Jim."

"I'll keep that in mind," Gene said as he lifted his glass. "In the meantime, I'd like to thank my friends who allow both Jim and me to enjoy the company of such a wonderful family. I just wish Julie and Seppa could be here also."

Aurora took the glass Marie was holding with a look that said *Pour yourself another one* and raised it, saying, "To the best friends and family anyone could ever hope for."

Chapter 5

As they began eating, Jim looked at Gene and asked, "Don't you think it's time to see if Marie is interested in your vacation proposal?"

Gene made a quick look toward Jim and then turned to smile at Marie saying, "What Jim's referring to is a trip I've asked him to take, and he agreed only if you could come with us."

Marie glanced from Gene to Jim and back, asking, "Is this something like that little vacation to Honolulu where you were gone for almost a week, and I got to come the day before you were coming home?"

"No," Jim answered, setting his glass down. "Gene wants me to go to Virginia and meet with some others who are working on an operation involving something to do with identifying illegal immigrants the government wants to deport. I told him that since I just got home, I wanted to spend the few days I have off with you and he said this way I could do both, meet with the other folks and spend time with you."

"When were you planning on telling me, and when would we be leaving?" Marie asked, laying down her fork and crossing her arms.

"I planned on telling you right now, and we would be leaving tomorrow morning," Jim told her, looking directly into her eyes. "I would have told you earlier, except I just found out an hour ago. As to the departure time, that's what I learned an hour ago also.

"But I can understand if you can't do it on such short notice, and I'll be sorry you can't go," Jim finished without breaking eye contact.

Marie looked at the determination in Jim's eyes for a moment and asked, "How long will we be gone?"

"I'll have both of you home the day after tomorrow," Gene answered. "This is just a preliminary meeting with a group of people who will be leading the effort to identify the subject groups and make arrangements for the government to round them up and transport them back to their homes."

"I thought your company only delt in security issues," Marie countered. "What does that have to do with illegal immigrants?"

"Normally, you'd be correct," Gene told her. "But due to the number of illegals we're dealing with, they've asked for our assistance because of their lack of enforcement personnel. Black Water has one of the most comprehensive libraries regarding facial recognition and probably the best communication monitoring systems outside of the NSA.

"Bottom line is that our government asked the company to assist them in one of the most massive deportation efforts ever seen," Gene continued. "And I want Jim involved because he has always been attentive to details and understands the logistics of such an operation."

"Then I guess you'll be needing him later when the operation starts," Marie stated. "More time away. What about his primary job, flying for American?"

"We can work around his schedule," Gene assured her. "And we still have some time before the operation begins."

"I've always thought your company did more than security," Marie told Gene. "That little coincidence where you identified the man who killed my husband, I guess that's part of the system of facial recognition you're referring to, isn't it?"

Gene looked at Jim and answered, "Yes. And we've used it before to help identify wanted people. Both here in the US and around the world. But it also helps us identify threats when we're safeguarding personnel we've been hired to protect."

"I think there's more to your *security* company than you're telling me," Marie said, looking pointedly at Jim. "Much more."

Jim looked at Gene and then admitted, "Yes. There's more. But I can't talk about that. You just have to trust me for now. However, this operation is exactly what Gene told you. We're assisting a branch of our government round up some very nasty people, and some very innocent children, and try to send them back where they came from."

"What children are you talking about?" Marie demanded. "I thought the whole deportation plan just involved criminals or other illegal immigrants."

"I'm sure you've heard about all of the children that the cartels were bringing into the country," Gene told her. "And most of them end up working in some very unsavory jobs, including child prostitution. We are just trying to get those kids back to their parents. The same for the people who just came here looking for a better life.

"We are just trying to help expedite their return," Gene finished. "Regardless of how these people got here, they have broken the law in one form or another, and we've been asked to assist to make the entire process as expeditious and fair as possible."

"What about the violent ones? Are you going after them?" Marie asked, looking at both of them. "What if Jim's injured? Or worse yet, killed?"

"Nothing is impossible," Gene answered. "But one of the things that I admire about Jim is his devotion to duty. And we both know that some things incur risks. You already know about how Jennifer died.

"We do everything humanly possible to take care of our people, but things can sometimes go wrong, just as with your husband David," Gene continued. "No, neither Jim nor I can guarantee nothing will happen. But we can't guarantee someone from some gang doesn't come in here some night and start shooting.

"Your husband knew that and swore an oath to protect, just as Jim did," he told her. "You've lost a husband to those people, just as Jim lost a wife and almost lost his life."

"But Jim's not under any duty now," Marie countered. "He served his time in the Marines and now he's just a civilian airline pilot."

Jim looked at Gene and then told her, "There are things I haven't told you. Now's not the time, but we can discuss some of it when we get to Virginia, if you still want to go."

Marie stared at Jim for a moment and finally said, "I'll go with you under one condition."

"What's that?" Jim asked, wondering if this was going to be his last meal with her.

"You tell me everything about your relationship with Gene and this Black Water company," she answered.

Jim glanced at Gene and, seeing him nod, told her, "Fair enough. I'll tell you as much as I can without getting into specifics. Just as David couldn't tell you everything. But you have to trust me and not push when I say I can't tell you something. It's not that I don't trust you. But some aspects of an operation aren't open for discussion."

"Did Jennifer know about this?" Marie asked.

"No," Jim answered.

"Did that stop her from getting killed?" Marie asked.

Jim sat quietly for a second and admitted, "No. But knowing wouldn't have prevented it either."

"Maybe I better not go tomorrow," Marie said, getting up. "I'm not sure how I feel about this right now. I'll call you after you get home, and we'll talk about it."

As she left the table, Jim said, "I think we better head home. Let her make up her mind as to what's more important to her. I'll go say goodbye to Tony and Aurora and take care of the bill."

"I'll get the bill," Gene said, rising and putting a hundred-dollar bill on the table. "Tell the folks that I'll be back in a couple of days when the dust settles."

Chapter 6

The next morning, after taking off from Dallas Love Field, Gene and Jim were sitting across from each other in the rear of the Gulfstream V as they headed east for Quantico. "What time is the meeting?" Jim asked as he picked up one of the breakfast bars from the airplane's galley.

"Noon," Gene replied. "I think most of the others got there yesterday and spent the night. I would have asked you to leave yesterday, but since it was so quick after your trip with American, I decided it would be better to give you an evening with Marie."

"Well, that didn't work so shit hot, did it?" Jim replied. "I never expected the reaction there at the restaurant, but I've determined that I don't have a clue as to what sets some women off. Even after all the time I've spent with her, that sort of surprised me."

"I'm certainly not the one to give advice when it comes to relationships with women," Gene said. "But I believe that if a man is putting something before her, whether or not it's a job, a ball game, or another woman, they react the same.

"I've tried to put myself in their place," Gene continued, "and I've come to the conclusion that in some cases, I'd feel slighted myself. Maybe not as much if they have a job that interferes with some plan I had, because I can understand that.

"I can understand if it's another person," he continued. "And I guess I understand when they feel you've been lying to them. Which, you must admit, you have. Maybe not an outright lie, but you and I have certainly shaded the truth."

"I know," Jim agreed as he rose to get the coffee pot from the galley to refresh their cups. "But it's not as if I'm free to discuss what I do for Muddy Water now or did for Dark Water when I was still in the Marine Reserves. My contract with Black Water is pretty clear on that issue."

"I believe there's some middle ground here," Gene told him as Jim refilled his coffee cup. "The company knows that there must be some sharing of the job within a family. A woman would never accept it if a man just simply said, 'I'll be gone for a week. Keep the house clean and the kids fed. Not sure exactly where I'll be or what I'll be doing, so just take care of things until I get home.'"

"I understand that," Jim replied. "But I've always been under the impression that we don't tell anyone exactly who we work for or what we do."

"Exceptions," Gene answered. "There are always exceptions. And, since you technically work for Black Water, and Black Water is not a covert company, there's definitely some leeway. Besides Black Water being an almost household name, it provides a massive umbrella for the international Dark Water and the domestic Muddy Water, and those two names are almost impossible to trace.

"They aren't publicly traded, nor are there any corporate entities that could be discovered. They are, in fact, nonentities. They don't exist outside of Black Water. Hell, even my name is only noted as an ex-military advisor. Your name is only discovered if a very close inspection of financial records is performed," Gene explained. "And even then, they would find an employee, or advisor, identified only by a company number that tied it to a bank routing number and its attached account number.

"If Marie is important, don't let the company's over-reaching security procedures prevent you from something you probably want well past your time with Black Water," Gene advised. "Now, if you have any doubts about what to discuss, or more appropriately, what not to discuss, don't hesitate to call me," Gene told him. "I can always put up a roadblock should it be necessary to derail any investigation. Or if necessary, I gave you permission.

"Now, if she's nothing more than a midnight mattress wrestling distraction, keep your lips sealed," he concluded. "Especially after you've won the wrestling match."

"You know the answer," Jim reminded him. "But I appreciate the advice, and roadblock should the need arise. Guess I'll figure it out based on her reaction when I get back."

"Good," Gene said, nodding. "Now, is there anything you need to know before the meeting tomorrow?"

"A shit load," Jim answered. "How many people are working with me? How many farms or strip malls am I covering? How wide are my targets spread? What sort of communication systems are we using? Do I have a dedicated asset back at Black Water? What weapons are being provided? Should I go on?"

"Not necessary," Gene responded. "Most of those things are common questions each leader for each area will be asking also. The briefing will cover most of your questions and will be based on data that has been gathered since I left a week ago.

"This entire operation was planned to take a maximum of four days, and because Muddy Water didn't have that many operatives available, Black Water assigned Dark Water assets," Gene explained. "Then with coordination with Homeland Security and ICE, we allocated them to areas where they could supply the maximum support to the overall deportation operation.

"And let's not forget that as far as the publicity generated, it will never show a Black Water asset. Every car, pickup, bus, or truck seen will belong to or be leased by Homeland Security. We will also be wearing black pants and either a Homeland Security jacket or an ICE jacket. Blackwater is not to be connected to this operation. Especially not Muddy Water, since they don't exist," Gene explained.

"What's my major role in the operation at Indianapolis?" Jim asked. "In the field or monitoring from some mobile location?"

"Monitoring," Gene answered. "Unless the shit hits the proverbial fan, then it's your discretion as to the best action to take and where you need to be."

"Can I divide my assets, or has that been done?" Jim asked. "And are there any of my team who I've worked with before?"

"Black Water made the initial assignments as to which location the people under you were to be assigned," Gene answered. "Most of the Dark Water people were assigned to targets involving TDA since we know they will be the ones most likely to put up a fight. Our thinking was that those folks had more training with operations involving foreign entities.

"Before you remind me that China is a foreign entity, we believe that most of them will put up little or no resistance. Especially the massage parlors, nail salons, and places where the girls will be found," Gene told him. "The farms TDA owns are a different issue. There will probably be some adult guardians, and they could prove to be troublesome.

"As far as the children from any location, we don't expect any trouble," Gene continued. "They may be hesitant since they've probably been told that the bad ol' US will lock them in prisons for the rest of their lives. But that should be the extent of it."

"What do we do regarding any guardians and with the farms?" Jim asked as he felt the plane start a shallow descent.

"Remove with as little disruption as possible," Gene answered. "You'll just make it more difficult to corral the children if you start a bloodbath and scare the shit out of them.

"Regarding the guardians at the other locations," Gene added. "They are mainly elderly women who started their lives in this country much as the little girls they are managing, and we don't expect any interference from them. Hell, they will probably help you put them on the buses knowing how their lives will end up if they stay.

"A final thought before we hang it up for the day," Gene said as the cockpit announced they were getting ready for the landing. "Most of these folks know we're coming. They have televisions. They've heard the incoming administration's scorched earth policies and know full well that the gates of hell are about to swing open to send every illegal home.

"We expect very little resistance on the Triad side," Gene said. "But the gates of hell are what we'll be opening with TDA. Those animals probably fear being sent back more than they do what they perceive as a pretty much toothless growl based on the previous four years. If there's going to be a bloodbath, it's going to be there."

Chapter 7

After landing at Marine Corps Air Facility Quantico, the Gulfstream taxied to the terminal set aside for visiting dignitaries and other civilian aviation that had been given permission to land there.

As soon as both engines were shut down, the stairs were lowered, and Jim followed Gene to where a black Suburban sat waiting. As they approached, the driver exited and held the rear driver side door open for him, saying, "Nice to see you again, General."

"You too, Carl," Gene replied, sliding into the car as Jim passed behind to the opposite side. "My friend Jim is accompanying me, and you'll probably be asked to bring him back here tomorrow afternoon for his trip home."

Carl glanced inside as Jim was shutting his door and said, "Mr. Lashley. So good to see you again, sir. Please don't hesitate to call me if there's anything I can do to assist you."

"Thanks, Carl," Jim replied, leaning forward to see him. "I'm sure the General and his crew will keep me pretty busy until I need to return, but thanks anyway."

"Anywhere we need to go before we hit the facility?" Carl asked, looking at Gene in the rearview mirror.

"Nope," Gene replied as they approached the double-fenced facility Black Water occupied on the base. "The guards should have our itinerary and access authorization, so just get us to the front door and we'll take it from there.

As they stopped short of the first gate, a black suited man with a clipboard approached as Carl lowered his window, saying, "General Barker and Mr. Jim Lashley on board."

The guard looked into the rear seat and nodded, saying, "Everything appears to be in order, please continue to the next gate."

As the twenty-foot gate slowly slid to the side clearing the road, a second guard appeared from the small shack between the two fences and waited until their car had stopped and the first gate slid closed behind them.

Nodding to Carl and the passengers, he glanced toward the grey cement structure that housed Black Water to ensure there were no vehicles approaching. He activated the second fence, which immediately began sliding to the side of the road, allowing them to enter the facility.

After Carl dropped them off at the main entrance to the building, Gene entered the code to open the door. Once inside with the door shut and locked, Gene turned to the guard inside a glass booth and said, "Just Jim Lashley and me today, Jerry."

As Gene held his Black Water ID to the window, Jerry looked from the ID to Gene and down to his list of permanent personnel allowed unfettered access to the interior. Satisfied that he had followed the procedures exactly, he pressed a button that buzzed and signaled the door into the interior was unlocked.

As Gene nodded his thanks, Jerry said, "Have a good day, General."

"You, too," Gene said, holding the door open for Jim.

As they headed down the central corridor that ran the length of the building, Gene suggested, "Why don't we grab something to eat before we get to the auditorium for the briefing?"

"Sounds good to me," Jim said, following Gene to the cafeteria. "That breakfast bar and ten cups of coffee won't hold me much longer. And if I know Debbie, since she doesn't always care to be called Bracer, she'll take the entire afternoon regaling us with the latest toy she's managed to develop. I swear, the lady is brilliant with computers and chips and apps and whatever things computer nerds find interesting, but she has the personality of an angry butcher whose beef cuts were just described as tough and inedible."

"I'll try not to let her ever hear about the angry butcher," Gene said, laughing as they headed for the food line. "But I must say it's pretty on spot. The lady can certainly be a little blunt. Grab what you want, and we'll take a seat over there by the back wall."

"Just blunt?" Jim asked as he preceded Gene through the line, selecting a salad plate, a ham and cheese sandwich, and a glass of unsweetened tea, and tossed a lemon slice on top. "The lady is so blunt that even calling a butter knife a scalpel would be a stretch in comparison."

Approaching the cashier, Gene waived his company credit card and said, "I've got his."

Once at the farthest table from the few other diners, Jim waited until Gene had taken his seat and then sat across the table. "I do have a suggestion, if you think we have time before the operation begins. By the way, just what has Black Water named this operation?"

"Black Water put all of their genius marketing and historical reference folks in a closed and locked room for almost a week before they announced their recommendation," Gene answered with a slight smile.

Jim waited a second and then asked, "And what was their recommendation?"

"Operation Deportation," Gene said, barely able to suppress his laugh.

"And the company signed off on that?" Jim asked, shaking his head.

"Of course," Gene answered, picking up his glass of tea. "It had never been used before, didn't really refer to anything specific, and actually made more sense than some oblique reference to some Greek God or obscure mathematical equation."

Still shaking his head at the notion of Black Water even using a name that could be construed as the actual operation, finally said, "Let me guess who originally made the suggestion."

"I know you're about to say Debbie," Gene said, nodding. "Now, why do you suspect her?"

"It's short, blunt, and to the point," Jim replied. "Typical of everything that woman does. No wasted words, no wasted time with anyone who can't stay up with her, and a rather subversive sense of humor."

"Guess that's our gal," Gene said, taking a bite of his tuna fish sandwich. As he glanced down at it, he asked, "Can you tell me why we call a tuna sandwich a tuna fish sandwich while we call a chicken sandwich just a plain old chicken sandwich? Shouldn't it be a chicken bird sandwich?"

Jim stared at Gene momentarily and finally said, "You've been hanging out with the wrong people for too long. It's starting to rub off."

"What do you mean? The wrong people?" Gene asked. "Hell, that's exactly your type of humor. But I guess I have been hanging around you for so long that it's rubbing off. Wonder if there's a vaccine for that?"

As they both shook their heads and laughed, Gene asked, "What was your suggestion, anyway?"

"I'm not sure if we have the time for the background checks, or if we even need them, but I believe this is exactly the sort of operation where we could use a couple of men we discussed before," Jim said, picking up his glass of tea.

"As far as time is concerned, we have four days from tomorrow to start our portion of Operation Deportation. And I know that it impacts your schedule with American, but I've got that covered and we'll discuss that later this afternoon after the meeting," Gene answered. "Just who are you referring to regarding this operation?"

"Butch North and Mike Knox," Jim answered.

"Aren't those the same two you were involved with during your little revenge operation after Jennifer was shot?" Gene asked.

"Yes, sir," Jim answered.

"And if I recall your answer when I asked if they should be looked at for additional personnel for Muddy Water, you said no," Gene continued. "And if I remember correctly, you said Mr. North was probably too easy going and you described Mr. Knox as a man who'd use a chainsaw when a scalpel would suffice."

"I stand by my assessment," Jim conceded. "But given this specific operation, I believe Mr. North would be a very nice point man involving the Triad portion, and Mr. Knox would be my first choice to head up the TDA side. Hell, I think Mike would enjoy it, and I believe Butch has exactly the compassion to use the delicate touch required with the little girls the Triad is exploiting and probably make them feel safer than a bunch of black suited men with chemical gear who come rushing into the massage parlors and nail salons."

Jim waited a few moments, letting the idea percolate within Gene's mind, and then asked, "If you agree, do we have enough time to do the background checks?"

"No problem there," Gene answered. "We did that when we thought you were going to use them when you wanted to eliminate the head of the group responsible for Jennifer's death.

"I only have one question for you," Gene continued. "Do you want the company to make them permanent operatives, such as you? Or do you see this is a one-time augmentation?"

Jim immediately answered, "For now, just a one-time augmentation. The company can then decide if they want them in future operations or not. But I'd feel better if I had a couple of people I know heading up the teams I'll be sending out.

"Basically, I'm a little concerned about being able to manage both operations, Triad and TDA, at the same time, given the extensive area and diverse style we need if we are to be successful without a major FUBAR and bloodbath to match," Jim concluded.

"When do you plan on approaching these guys?" Gene asked as he saw the obvious look of concern on Jim's face.

"As soon as I get home," Jim answered. "If you and the company agree with me and my reasoning, I'll call each of them before I leave here tomorrow and set up a meeting.

"The only thing I need from the company is basically the same package you gave me when you recruited me," Jim continued. "And I'll handle everything else, including making sure they understand the ramifications of the job and the obvious restrictions it entails."

"I'll get the personnel folks working on it while we attend the briefing from Bracer and should have you an answer by this afternoon," Gene said checking the time. "But now we need to get to the auditorium or suffer the wrath of Betty Blunt."

Jim laughed as he carried his tray to the area for pickup and said, "Guess I shouldn't ever use that phrase in her presence."

"Not unless you think life as a eunuch would be desirable," Gene said, setting his tray down.

Chapter 8

As they took their seats near the rear, Debbie approached the podium and announced, "Folks, if you could please find your seats, and by that I'm referring to those standing in the rear, we'll see if we can get out of here in the next couple of hours."

As the few people quickly found an empty seat, she hit a button that illuminated the large screen behind her and said, "Here is a quick overview of where each team will be deployed. There will be a packet at the rear that has more extensive details regarding your specific area. Pick it up on your way out.

"Black Water has spent almost a year monitoring the people we're interested in and their details as well as a photo of each person we're looking for," she continued. And by photo, I'm talking thousands that have been reduced to a single frontal and side-view photo."

Switching the screen to another photo depicting a pair of normal glasses, she announced, "Each of you will be given a pair of these glasses. They are a new development that incorporates a minute camera, a tiny screen in the upper right corner of the right glass lens, an almost invisible thread-sized cable connected to an earpiece, and a microphone that will pick up your voice even if it's merely a whisper.

"The purpose of the glasses is to allow you to look at any individual, focus on them for more than one and a half seconds, and the glasses will send the person's face to the app we have on the phone you will also be given," she continued.

"Now, the photo will be immediately sent to our data bank here at Black Water headquarters, and it will be compared with our facial recognition program that is one of the most extensive in the world," she explained. "It will be almost instantaneous since it first compares it to the special file we've built using only the people whom we've identified as part of the TDA or Triad targets.

"As soon as recognition is achieved, it will confirm the target and his, or her, photo from our file will be displayed on the right lens as I described," she added. "Now, this allows you to see a potential target, hold them in view for a second and a half, and receive confirmation within the next second," she said, smiling. "I've been working on this system for quite some time and am positive it will allow each of you to feel confident that you're looking at a valid target."

She waited for a moment to allow the magnitude of this system of target recognition to sink in, and continued, "Now the phones we are providing are also equipped with an enhanced microphone that will pick up the slightest sound within fifty feet. It will be transmitted, run through a noise filtering system, and retransmitted to the earpiece built into the glasses.

"For the team leaders, you will have an iPad that will display your entire area of responsibility and will show the exact location of each member within three feet of where they are standing, or sitting," she told them. "It will also show any of the targets we've identified over the last year.

"For those of you who haven't worked with this system before, we've managed to put a bug into the phones they are using. It will both pick up any conversation they are having with anyone within a fifty-foot range, relay it to us, and also provide a location within the same three-foot criteria as your phones provide.

"This should make it easy for the leaders to see the entire operation of his area and zoom in on any specific area or person by merely tapping the screen of the iPad he will be given," she continued. "Pretty much standard for a touch screen, the area can be compressed or expanded by pinching or opening two fingers. You can also use the screen to select any individual for a voice connection by holding your finger on the figure representing that specific person.

"The screens will also show any member of Black Water, be they Dark Water or Muddy Water, with a green icon," she told them. "Any target will be shown in red, and any unknown will be a white icon. If any member wants to make an assessment of a particular individual, using the method already described will relay his photo and we can make an assessment.

"We don't anticipate having to use this since we've compiled such an extensive file on all of the known targets," she continued. "But there is always a chance that an unknown but viable target enters the area. We have an extensive file on certain identifying characteristics, such as tattoos, that assist in recognizing members of either TDA or the Triad.

"If identified as such, his icon on the iPad will change from white to blinking red to alert the leader, and he will make the determination as to the disposition of that individual," she added.

"Our contract only requires us to gather the targets and turn them over to the appropriate entity," she said as the screen changed to show several buses and vehicles. "One or more of these vehicles will be either at or near the locations, such as the massage parlors or farms, and can be sent in by the team leaders when the operation requires removal of the targets.

"Should additional vehicles be required, there are a limited number at the command vehicle's location and an equal number of drivers.

"Please note that either Homeland Security or ICE is on the side of each," Debbie told them. "There will also be an ambulance near each of the locations in the event someone is injured. If it's one of the targets, they will be taken to certain selected medical facilities. And, before you ask, the ambulance personnel are trained EMTs working for us.

"Should any target be exterminated, leave them behind and notify your leader," she explained. "He will either use one of the vehicles you arrived in or the ambulance to remove the body and transport it to a designated location where it will be disposed of, which is primarily handed over to the government agency.

"Now, each member of the teams will wear the uniform associated with the government branch as represented by the vehicle," she told them, switching photos on the screen again. "You will be issued weapons and other items we believe necessary for your part. This includes such items as flex cuffs, duct tape, blankets, hoods, tasers… pretty much anything we could think of to cover any contingencies.

"One additional issue that we've just learned," Debbie told them. "The FBI has now joined the operation on the government's behalf. Now, I shouldn't need to say this, but please don't go pissing off either them or the other government agencies. First, we're trying to be relatively obscure. The company wants to remain one hundred percent invisible to either the public or any of the agencies that are tasked with the deportation.

"Avoid any interaction with any of these agencies," she directed. "If it appears that some contact is imminent, let your team leader know and let him give you some advice as to how to handle the situation."

Debbie glanced at her watch and then asked, "Are there any questions?"

Seeing a hand go up in the rear of the auditorium, she said, "Yes? What is it that you don't understand?"

"How do you know what size of uniforms everyone needs?" the man asked. "I'm the team leader and I don't even know everyone I'll have working with me."

"Let me see if I can accurately explain the elusive facts," Debbie answered as she looked down at the podium for a moment. "We, Black Water, know every single person assigned to this operation, and that includes you, Mr. Sam Johnson. And I know that you wear a size forty-two jacket, thirty-six waist pants, and nine and a half size loafers, which you favor. Does that help you understand that this company doesn't leave much to chance? By the way, you should really work on your pant size. But I'm sure there isn't time between now and the onset of the operation for us to supply you with a smaller size."

In the rear of the auditorium, Jim leaned over to Gene and whispered, "Have none of these people ever attended one of Bracer's briefings? If they had, they would know to just sit quietly and keep their mouths shut."

"Unfortunately, there's always some fool who thinks he can catch her unprepared," Gene whispered back. "There's always one or two who try to impress the rest of us with their intelligence and a grasp of some miniscule point."

Just as Gene finished, another hand went up from the middle of the auditorium. "Yes?" Debbie asked, looking at the man.

"You mentioned how we were to use the ambulances to take any injured or deceased target to certain designated medical facilities," he stated. "What about if any from our team is injured?"

Debbie once again glanced down at the podium and responded, "Well, Mr. Raymond Reynolds. The company has decided that every one of you has volunteered for this operation and should have known the risks. Should you become injured or if one of your team does, just use the phone we've provided and call 911."

Debbie paused looking at the man and then said, "You've got to be shitting me with such a question. Now, are there any relevant questions that don't border on the absurd? If not, this meeting is over."

As she started to turn, she stopped and said, "I guess I need to be a little more direct on a couple of things, stop at the rear and get the glasses and phones along with the instruction pamphlet before you leave. Then make sure you understand their operations so we can ensure they are functioning properly. And yes, Mr. Reynolds and Mr. Johnson, the phones can still be used to call someone should you have any further questions.

"And oh, by the way. The reason I could determine exactly who was questioning is because of the effectiveness of my facial recognition and precise location program. I have it on the iPad here on the podium. If that doesn't give you confidence in the effectiveness of the services we can provide you out in the field, you need to seek employment as a greeter at your nearest Walmart. And I believe you'll fit right in there, Mr. Johnson," Debbie said, turning from the podium as the screen behind her went dark.

Chapter 9

"Jim Lashley, it's been a long time," Debbie said as she caught up with him and Gene at the rear of the auditorium. "What's new with my favorite operative?"

Jim gave her a quick hug and replied, "If I'm your favorite operative, why is it you never call on me during one of these meetings?"

"Because you're smart enough to keep your mouth shut, unless it is really essential," she told him. "I can't believe some of the crap that comes out when these folks are supposed to be professionals."

"Let me see if I can shed some light on it," Jim said, smiling. "You're familiar with the Bell Curve, so what portion of the intelligence curve is to the left of the median line?"

"One half," she answered. "I suppose you are going to remind me that means half of the population is to the left, or more succinctly, below average intelligence."

"Bingo," Jim said. "You just won a cupie doll."

"I realize what you are saying, but I would have expected Black Water to have a slightly higher standard than average when it comes to hiring," she retorted.

"We do our best," Gene told her. "But sometimes you can't tell if the egg is rotten until you crack the shell."

"Aren't you old enough to know what 'candling an egg' was?" Debbie asked. "I'll rephrase since I don't want to demean anyone who has advanced through life more than I.

"Never mind the question," she continued. "Candling an egg was holding up against a lit candle and inspecting the contents to see if it contained an embryo or not. I guess there's no such thing as that to see what's happening in the shell we refer to as a skull. Maybe an otoscope?"

"The light isn't bright enough," Jim told her. "But I do have a request that I'm sure you can handle. I need two more phones, iPads, and glasses for a couple of new recruits. And I'd like them paired with my phone and iPad."

"Why this last-minute change?" Debbie asked, looking from Jim to Gene. "I've already met the tasking for this operation, if I'm not mistaken."

"Jim has asked for two more men in order to cover the area he's been assigned," Gene told her. "I support his request and would appreciate it if you could have the phones and glasses ready before he leaves in a couple of hours."

"Not a problem," Debbie said, nodding. "Do you need the same uniforms the others will be using?"

"Yes," Jim told her. "I'll call back with the sizes as soon as I confirm they will be working with me."

"I'll take care of it when you let me know," Debbie replied. "The uniforms and other items will be in Indianapolis when you get there. Anything else?"

"No, that about covers it," Jim answered. "I'll work with the new guys to make sure they understand how the glasses and phones work. We'll be calling you in a couple of days to make sure everything is good on our end."

"Don't wait too long," Debbie reminded him. "You need to have them ready to go in five days, if I'm not mistaken."

"You're never mistaken," Jim said, smiling. "That's one of the things I've learned about you. I'm heading home this afternoon, and I plan on contacting both men tomorrow, unless one of them is on a trip."

"If you're referring to North, he's not on the schedule," Gene told him. "I took the opportunity to check with the schedulers and crew tracking while we were enjoying Debbie's presentation."

"That leaves Knox," Jim said. "I'll call them as soon as we're done here."

"Unless you have anything else, I'll go get the glasses, iPads, phones, and be back here in ten minutes," Debbie said.

"Meet us in the cafeteria, if you don't mind," Gene told her. "I want a glass of tea and a few minutes to talk to Jim."

"I'll be there in ten," Debbie responded, turning away.

"Let's go grab a table," Gene said, walking out of the auditorium. "There shouldn't be too many people there right now, and we can check out the glasses, iPads, and phones when Debbie gets there."

After taking glasses of unsweetened tea with lemon, Gene led Jim to a rear table and sat asking, "Are you confident Butch and Mike will fit with your operation?" I know you've told me what you planned, but I have some concerns about the TDA."

"What's your concern?" Jim asked, taking a seat across from Gene.

"I had some quick research done while I was checking on North's flying schedule," Gene answered. "By the way, he did have a conflict with the days of the operation, but his trip was suddenly needed for a Captain's upgrade."

"Strange how that always happens," Jim said, shaking his head. "And I suppose my trip has already been reassigned, as well."

"Matter of fact, it has," Gene replied, smiling. "There are a lot of training requirements at American this month."

"Understood, now what did you find on Knox that causes you concern?" Jim asked as they saw Debbie heading their way.

"Apparently, he was charged with murder while stationed in Vietnam," Gene answered. "He was left to guard some prisoners and basically shot every one of them.

"The only reason he wasn't found guilty at his court-martial was that the orders he received weren't specific," Gene continued. "The order was to make sure none of the prisoners escaped. Period.

"And Knox's attorney made the argument that he followed the orders because none of the prisoners did escape," Gene finished. "I want you to make sure any orders you give him are clear and leave no doubt as to the intent of the orders, and the limit to the scope of his actions."

"Are you asking me to tell him not to kill any of the targets he encounters?" Jim asked as Debbie placed the glasses, phones, and iPads on the table.

"No, I just want you to make sure he knows to use a little restraint unless drastic measures are required," Gene answered. "Make sure that *guard the prisoners* means to retain them, not to destroy them. We would like to have as many as possible to get as much information about their activities both in the US and other countries. We really want to find who's controlling them.

"We're pretty sure Hector Rusthenford Guerrero Flores, alias Nino Guerrero, is still the leader, but we would like to find out exactly where he is since his escape from the Tocoron prison, and if he's still in charge," Gene explained. "This is important. These folks are like MS13 on steroids and our government would like to cut the head off the snake before it expands any further into our country."

"I'll make sure Mike understands the limits," Jim assured Gene.

"Sounds like you've got a tiger by the tail," Debbie observed as she handed Jim the phones. "Maybe I should put a little taser into the phone so you can emphasize the limitations of going rogue."

"I'm sure he'll understand after we talk," Jim replied. "You've also got to remember he was just a teenager when he was tossed into the jungles with little more than a basic set of rules meant more for survival than anything else. Don't forget that I was there, also. It's just fortunate I didn't have to make the decision he was forced to make.

"And let's not forget we don't know what prompted him to annihilate the prisoners," Jim concluded. "Maybe there was a perceived threat. Hell, I wasn't there, and neither were any of us. Put yourself in the boots of a scared teenager who has just watched more than a couple of his friends blown to pieces by these same people. Then tell me what you would have done. Remember, Marines were taught to kill the enemy. Not much training regarding guarding prisoners was ever given."

"Got it," Debbie said as she set the iPads in front of Jim. "These are pretty much automatic, just turn them on and they are married to the phones and that portion of the operation.

"These are the iPads your new guys will have, and they mirror yours," she continued. "The phones are pretty much the same. The glasses are the same as all the others. If you can operate yours, which I'm sure you can, you can show them how to use them. If they have any issues once in the field, we are monitoring their phones. All they need to do is say what they need, and we'll hear them. No need to make a call, we're listening."

"I'll make sure they are up to speed," Jim said. "And, thanks for everything you've done here, Debbie. This is absolutely terrific. Tops what we had in El Paso, and that was superb."

"Just taking care of business," Debbie replied. "And that business is keeping my boys safe."

"Even Reynolds and Johnson?" Gene asked, smiling as Jim gathered all the items.

"What's the old saying?" Debbie asked as she walked with them to the door. "The Lord protects fools and idiots? Well, not to underestimate his reach, I feel as if I'm lending a hand with these two. And even the two of us might be struggling to protect their level of stupidity."

Chapter 10

Early the following morning, Jim headed for Aurora, Texas, to meet Mike and Butch at Butch's Check 6 Ranch. As he arrived and headed up the drive through the open gate, he noticed several black Angus cattle grazing in the pastures on either side of the asphalt driveway just outside the white pipe fence.

Arriving at the stucco ranch-style house, he saw an old dust-covered red pickup sitting on the driveway just short of the garage. Pulling beside it, he was met by both men as he stepped from his 'Vette.

"Long time," Butch said as he walked up and shook Jim's hand. "Mike and I have been discussing exactly why you wanted to have this meeting that you described as urgent."

"We'll get into that after we go over a few administrative issues," Jim replied as he pulled a briefcase from the passenger seat of the car. "Could we go inside where there'll be room to spread out some papers and see if both of you want to get involved?"

"No problem," Butch said as Jim shook Mike's hand. "How about we go to the porch and see what you think is so important?"

"How's life up there in Vashti?" Jim asked Mike as they followed Butch to the rear of the house.

"Generally quiet," Mike replied. "Just me, the cattle, and a chupacabra or two."

Jim laughed as Butch opened the screen door leading into the porch. "I didn't know you had such an infestation of the rare and elusive chupacabra up there. I thought most of them were down around the Rio Grande just outside of Del Rio."

"Migratory little beasts, they are," Mike told him as Butch pulled out a wrought-iron chair and motioned for them to have a seat.

"I've got a pitcher of tea, if you're interested," Butch offered before sitting down.

"Maybe later," Jim answered, opening his briefcase. "Maybe we'll need something with a little more kick after we finish this little discussion."

As Butch and Mike took seats, he started, "I'm sure you both remember the little issue involving the improvised explosive devices a couple of years ago. And the fellow Mike caught on his property."

As both nodded their agreement, Jim continued, "What I didn't tell you at the time was that I work for an organization called Black Water, and unfortunately, your names came up when I was trying to take care of some rather personal business involving the death of my wife."

"Is that the reason you were interested in the feral hogs I have running rampant on my place?" Mike asked.

"Slight indiscretion on my part," Jim answered. "I went a little overboard with my desire for revenge. But let's skip that and get to the point.

"Black Water, as you may know, is an organization that contracts with our government when something needs to be taken care of that would not be looked upon with favor from the masses," Jim said as he handed copies of the standard Black Water agreements to both of them.

"Before I get into the specifics of some of our contracts, or operations, I need both of you to read these contracts, ask any questions, and agree before I say much more," he finished as Butch and Mike picked up their copies.

Waiting for them to read the details, Jim sat back and studied both of them.

As Mike read, he asked, "Is Black Water part of the Central Intelligence Agency?"

"No, not exactly," Jim explained. "They are more like Air America was over in Vietnam. They were contracted by the CIA to provide certain services that the government couldn't be involved with."

"Sounds sort of like Abu Ghraib," Butch said as he laid his copy on the table.

"*Sort of* is a good way to phrase it," Jim agreed, nodding. "There are issues all over the world that require operations that none of the governmental agencies can perform. That's not to say they can't, but there has to be reasonable deniability for our government. That's where Black Water comes in."

"So, we're not technically working for the government, but being paid by them to take care of the dirty laundry," Mike mused, setting his copy down.

"No, if you accept the conditions spelled out in the contracts, Black Water will be paying you," Jim clarified. "I don't know exactly how the money is transferred from the government, but it's probably for some study on the environmental impact of the chupacabra on domestic ranching operations. Or something just as obscure. And we contract for services with other entities besides our government. However, now is not the time to be concerned with who pays whom.

"Let's not get ahead of ourselves," Jim added. "I've said about all I can say until both of you sign the contracts. But make sure you fully understand that any breach of the contract could have severe repercussions.

"So, ask now if you have any questions," Jim finished. "Once you sign, if you sign, it goes immediately into effect."

"There isn't any specific time mentioned," Mike observed. "My only previous contract with the government was when I joined the Marines, and that was for a specific period of time. How long is this contract in effect?"

"Pretty much indefinitely," Jim answered. "Even if you aren't an active member of Black Water, everything you did or learned is what you might think of as covered by a nondisclosure agreement.

"For now," Jim explained, "I'm offering you a specific operation for a specific time frame. It's up to the company if further employment is offered.

"And, once you sign, it doesn't mean that you have to accept any assignment," Jim explained. "This is where Black Water and the military differ. In the Marines, you were given an order and expected to follow it. Here, an operation may be offered, but you can pass if you don't feel comfortable with it."

Butch looked from Mike to Jim and asked, "Does that include whatever it is you've come out here to offer?"

"It does," Jim confirmed.

"So, if I sign and don't want any part of it, I'm free to refuse?" Butch asked.

"Certainly," Jim answered. "But the nondisclosure part is still in effect … for the rest of your life."

"Hell, this can't be worse than Vietnam," Mike said, picking up the pen Jim had laid on the table. "I'm in."

"Can you tell me anything about the operation?" Butch asked holding the pen Mike handed him.

"No," Jim answered wondering if he had misjudged Butch. "Not until you sign the agreement."

Butch sat back looking at Jim and finally said, "You probably saved my life when you notified me of the IED on my gate. And since you said I could refuse once I hear it, I guess I'll at least listen."

As Butch finished signing his contract and passed it with Mike's to Jim, he asked, "Just what does this involve and when do we get involved?"

Jim looked at both signatures, put the contracts in his briefcase, and asked them, "What do either of you know about the Chinese Triad or Tren de Aragua?"

Chapter 11

For the next hour, Jim explained what the problems with both groups were and how they fit into the government's deportation plan for the illegal aliens who had breached our borders.

He then went into great depth regarding the human trafficking conducted by the Triad and the violence by TDA. When he went in-depth regarding the operations involving children by the Triad and the rape, murder, and other violent acts by TDA, he waited for their responses.

"Son-of-a-bitch," Mike said venomously, I've heard about some of the stuff TDA was doing across the country, but I didn't know about the sex stuff involving kids. Son-of-a-bitch. Those assholes need to be fed to the hogs."

"Not enough hogs or time," Jim told him. "We have three days to begin this operation. Your involvement will be conducted in Indianapolis, and we need to be there two days from now. But before I get into that, is there a chance you have any Jack Daniel's handy?"

"Be right back," Butch said heading into the house. "Need any chaser?"

"Just the Jack for me," Jim answered.

"Couple of ice cubes," Mike told him.

As Butch returned with three old fashion glasses and a bowl of ice cubes, he asked, "How much information do we have regarding this operation?"

"Extensive," Jim answered, taking one of the glasses and pouring in a couple of inches. Handing the glass to Butch, he continued, "The company has been researching the targets for several months. And when I say researching, I mean down to the color of their shorts, bowel movements, or any other aspect of their miserable lives."

Pouring another glass and handing it to Mike, he continued, "I've never known any organization that has as much technological knowledge as Black Water. They make the NSA or CIA seem like kindergarten compared to a PhD as far as intelligence gathering capabilities.

Pouring a glass for himself, he raised it and said, "To a successful operation and extermination of some of the worst people to ever cross our border. Welcome to the dark side of life."

After each had taken a sip, Jim set his glass down and stood, saying, "I'll be right back with some of the equipment you'll be using."

A couple of minutes later, he returned and handed both of them the phones, glasses, and iPads. Waiting for them to pick them up and examine them, he explained, "The iPads are synced to mine and the company. I'll get into more detail after I go over the other items.

"The phones are connected to a staff of folks back at Quantico who will be monitoring your every move or conversation," he explained. "The phone will pick up any noise or conversation within about fifty feet. It will also provide your location within three feet, and that location will be displayed on the iPad, as well as on mine and those back at Quantico.

"I'll get more into how to use both later, but if you'll put the glasses on, I'll demonstrate how they work," Jim said as he took out his phone.

"Debbie," he said as the phone was answered. "I'm here with our new recruits and I'd like for you to provide a small demonstration, much as you did earlier."

"No problem," she answered. "Do they have their phones and iPads on?"

"Hang on a second," Jim told her as he passed on the requests.

"They are on now," Jim said. "I'll put mine on speaker so they can hear you."

"Gentlemen," Debbie said, "welcome to Black Water. Would you please put on the glasses, state your name, starting with Mr. North, and look at your iPad."

"Butch North," Butch answered holding the phone up to his face.

"No need to even hold the phone, Mr. North," Debbie told him. "Just lay it on the table I see there and look at Jim."

Waiting until Butch was looking at Jim, she continued, "You'll see a small picture of Jim in the upper right corner of the right lens. That's where you'll see anyone or anything you're looking at. And I can see it also. Do you see him in your glasses?"

Hearing that he did, she then said, "Now, look at your iPad. You'll notice a green icon representing Jim, another one for Mr. Knox, and one for yourself."

Pausing to give him time to look at his iPad, she asked, "Does this represent the seating arrangement and location of each of you?"

Again, hearing he did, she continued, "I'm going to send a picture to demonstrate the other aspects of the glasses. Let me know what you see."

"Osama Bin Laden," Butch replied as the picture appeared in his glasses. "Well, I'll be damned. Pretty neat."

"Now, Mr. Knox, your turn," she said. "Since you've heard and possibly seen most of the capabilities of the equipment, make sure your phone and iPad are on and look at Jim."

"I see him," Mike said. "And I see a picture of him right where you said it would be."

"Look at the iPad," Debbie directed. "See the green icons I discussed with Mr. North?" "Yeah, I see them," Mike acknowledged.

"Now, Jim, please get up and walk away from the table," Debbie directed.

As Jim approached the screen door, she asked, "Do both of you see how the icon follows Jim?"

Both answered yes, and she continued, "Any of our targets will also appear as red icons if they are within about fifty feet of you. Any unknown will be white. But all you have to do is look at them the same way, and we'll identify the person, and Jim will then determine what action to take with them. Are there any questions?"

"Do I get to keep this stuff when the operation is over?" Mike asked. "I'd like to wear the glasses next time I go to the bar. It would be nice to identify everyone in there before I make a move on some lady whose husband is standing beside her."

"You bet, Mike. The company has unlimited resources to help you find the woman of your dreams," Debbie said sarcastically. "At least we could give you a little information about her, so you don't make the same stupid mistakes you've made with the last two."

"Just how much do you folks know about us?" Mike asked, taking off the glasses.

"We know more than you can probably remember, given your propensity for alcohol," Debbie answered. "And I can probably guess pretty accurately about what size shirt and pants both of you need for the uniforms you'll be wearing during the operation, but since I can't see your shoes, I'd appreciate if you'd provide me with all three … shirt, pants, and shoes."

A few minutes later when both Mike and Butch were satisfied with their equipment, Debbie said goodbye and Jim refilled their glasses asking, "Have you ever seen anything like that?"

"Hell, I've never even heard of something like that," Mike answered, taking his glass. "Friggin' amazing."

"No shit," Butch responded. "How the hell does she know that much about us? We just signed the papers a couple of minutes ago."

"Background research on prospective operatives," Jim answered. "And she knows more about me than I wish. She's good. Very good. But just beware, she's also lacking in what some may call social graces."

"Sort of like that comment about my two ex-wives," Mike said, shaking his head. "Or my taste for the finer whiskeys. I damn sure wouldn't want her watching every move I make."

"Me neither," Butch agreed. "Please tell me that my private life will remain private. Not that I'm very interesting, but crap … I don't want to go through life wondering who's watching."

"Not to worry," Jim assured them. "I've had an issue or two before with their ability to step on my toes, and it's been resolved.

"Now, I've got some personal business to attend to, so if there are no further questions, I'll be giving you a call in the next day or two as to when and where to meet me, and we'll head for Indianapolis."

"What about my schedule with American?" Butch asked getting up. "I've got a trip coming up in a couple of days."

"You'll be getting a call from crew scheduling, probably tomorrow, about how they need your trip for some training bullshit," Jim answered. "And yes, they know just about everything that involves your schedule when it conflicts with their needs. And they always resolve it in their favor. I'll be in touch. Please take care of the equipment and have it with you when we meet again."

Chapter 12

As Jim headed home, he tried calling Marie … twice. No answer. Just as he was going to try again, the phone rang.

Thinking it may be her calling, he quickly answered, "Hey, how's it going?"

"Not so good, but so nice of you to ask," came the answer from Gene. "How'd your visit with North and Knox go?"

"Good," Jim said, heading east on Highway 114. "I have their contracts with me, and they have a pretty good understanding of the equipment. I haven't told them very much about the operation in Indianapolis, but they seem to be ready.

"Speaking of ready, when do you need us in place?" Jim asked, slowing for the inevitable traffic at the Texas Motor Speedway and intersection with I-35W.

"I'll have a plane at Love Field in two days, should be there ready to take the three of you to Indy at eleven o'clock," Gene answered.

"I'll give them a call when I get home and make sure they are ready," Jim replied, passing the ever-increasing number of developments along 114. "I think I may have them come to my house the night before, so we'll all be there when the plane arrives."

"I thought you might be having Marie over," Gene mused. "How's that going?"

"Don't know," Jim answered. "If I had to guess how it's going, I'd say for now it's pretty much going to hell."

"Oh? What's the problem?" Gene asked.

"Beats the hell out of me," Jim replied, entering Westlake. "I've tried calling, and she's either not answering or her phone was stolen. But I'll deal with that after this operation is over.

"Now, what did you mean about the operation not going so good?" Jim asked. "Do we have a problem?"

"Yes and no," Gene answered. "Don't know if you've been keeping up on the news, but a crowd of thousands tried to sneak up on a few TDA members in Colorado yesterday.

"There were units from the FBI, CIA, ICE, DEA, Homeland Security, City Police, County Sheriff's office, Customs and Border Patrol, hell, I think there may have even been a troop or two of Boy Scouts," Gene exclaimed.

"Needless to say, with so many knowing what was going to happen, word of the raid got out and there wasn't a single TDA member in the apartments," Gene added. "Hell, whoever thought up this raid using all those organizations should be fired. You can't have that many people knowing about a *surprise* raid and expect it to be a surprise.

"Worse yet, we weren't notified," Gene continued. "We're supposed to be a part of this overall deportation, and those idiots pull a raid without informing us.

"Could have been worse, I guess," he added. "We could have been in the middle of an operation, and some small-town Barney Fife city deputy would have pulled his single bullet out of his shirt pocket and taken a shot at one of our people.

"Let's not forget that we had eyes on that group of TDA and could have let them know what was happening before they made fools of themselves on national TV," Gene concluded. "What a FUBAR shitstorm."

"Is that the yes part of the problem, or in the 'no' column?" Jim asked, thinking about the possibility of the failed Colorado operation impacting his in Indianapolis.

"I'm going to put in the no problem column," Gene answered. "We still have eyes and ears of the TDA members who evaded the disastrous public display of incompetence. We know most of them headed for Chicago or, lucky for you, Indianapolis.

"There were about one hundred members in Colorado, and it looks like forty of them are now in your area," Gene continued. "However, I don't think it will have any significant impact on the operation there. I'll be sending half of the folks that were assigned to the Denver operation to you and the others to Chicago.

"I've met with the upper folks at Black Water and expressed my opinion, but there's nothing that can be done to undo what happened," Gene finished. "Not sure if they can control those government organizations or not. Those folks seem to be more interested in seeing themselves on the news than taking care of the problem. It's gotten to be too damn political."

"That seems to be the problem most of the time," Jim responded as he entered Trophy Club. "Some Senator or Congressman sees a chance to appear aggressive regarding some issue, and he'll blow a solid operation with a single speech or article in his local newspaper."

"True," Gene agreed. "But this may also lead the TDA into thinking they'll find out about any deportation operations before they happen. Especially since there are hundreds of morons who are standing in front of every camera available and decrying the inhumanity of deporting those who have been raping, murdering, or robbing their neighbors."

"As much as I appreciate all the features of our cell phones, they can be disastrous when trying to conduct operations," Jim replied. "One of my concerns since this deportation issue has garnered such opposition, is having our people exposed by some bystander with a different point of view to video an operation."

"I know, and we're taking as many precautions as possible," Gene assured him. "That's why everyone will be wearing the appropriate uniform, and along with face shields, we're hoping that won't be a major problem."

"Speaking of that, I'm assuming that in addition to having the folks who are going into the fields or massage parlors outfitted with the uniforms, the team leaders, such as I, will have the same if we need to leave our mobile headquarters," Jim said slowing for the traffic that was merging into the single right lane.

"Absolutely," Gene answered. "And I strongly suggest you wear it regardless of where you are once we begin. Even if you're not involved with a specific action, you could be seen and that could lead to speculation about rogue operations."

"Understand," Jim told him as he crept by an assortment of flashing red and blue lights beside two cars that obviously tried to occupy the same space at the same time. "Anyway, I've got to go get some groceries to feed Mike and Butch tomorrow evening."

"I guess you're not taking them to Siciliano's," Gene responded, chuckling.

"Nope," Jim answered. "Not sure I'm welcome there right now. But I'll try one more time to contact Marie this evening. If she doesn't answer or return my calls, there's not much more I can do."

"Don't give up on her yet," Gene advised. "I think she'll understand the subterfuge you thought was necessary and see it wasn't some scheme to keep her from knowing your involvement in our operations.

"She's a bright lady," Gene continued. "It's no wonder that she put two and two together and knew that when you said it was three or five, something was amiss."

"I think that's part of the problem," Jim told him. "She may feel that since David was killed on an operation, she either didn't know about or didn't understand, she's afraid I'll be going off and not coming back like him. As long as she just saw me as some ordinary schmo pilot, she didn't worry.

"But I don't have time to do much about it right now," Jim finished as he pulled into his driveway. "There'll be plenty of time when this little deportation thing is complete."

"I agree," Gene said. "Now's the time to focus on the mission. I'll see you in Indianapolis when you get there in a couple of days."

"You're coming to Indy?" Jim asked as the garage door rose.

"I'll be making a visit with each of the teams," Gene answered. "You know I've never been a behind-the-desk Commander. I want to make sure everyone involved knows I'm behind them if they need anything. Anyway, let me know if you think of anything before you guys get to Indy."

Chapter 13

Once in his house, Jim tried again to contact Marie, and didn't get an answer. Next, he called Butch and told him of his plan for them to spend the night at his house the night before the flight to Indy.

As Butch answered, he asked, "Still interested in the operation?"

"So far," Butch answered. "Mike and I discussed it after you left, and we both decided that we would make our final decision when we get to Indianapolis and see exactly what we're getting into. Would that be too late?"

"No," Jim answered. "But you'll leave me with few options. And I can guarantee the company will never offer employment again."

"I figured that," Butch replied. "Neither of us wants to back out since giving our word, but I, more than Mike, have a few concerns regarding how dangerous this could turn out to be."

"I can't guarantee there won't be any danger, but I can assure you that the company doesn't put any of its people at risk unnecessarily," Jim offered. "I've been with them for several years, and this operation is one of the most benign that I have been assigned.

"If it puts you more at ease, I'm putting you in charge of the Triad portion," Jim continued. "I really don't expect any resistance from the ladies at the massage parlors or other business the Triad is operating."

"How many of these businesses are there?" Butch asked.

"I don't know yet," Jim answered. "I may get more information before we leave, but it may not be until we're there that we will be given the targets. And you'll primarily be directing others who will be doing the actual work in getting the people onto the buses.

"Between the intelligence you'll be getting from Black Water and the protective gear, I'd say the risk is minimal," Jim assured him. "If everything goes as planned, you will probably remain in the van with all of the communication gear and just direct the teams."

"What will I be directing them to do?" Butch asked.

"Pretty much just making sure they have the buses and vans to accommodate all of the targets, possibly reassigning teams or some of their members to a different location, if necessary," Jim replied. "There is always a chance you'll be required to take a more active role, but with the protective gear and our ability to keep you informed of any threat, I just don't see much chance of you being injured."

"If I'm going after the people from the Triad, I guess you're assigning Mike to the TDA," Butch said.

"That's correct," Jim told him. "Mike has the experience from his two tours in Vietnam to assess the situations most likely encountered with TDA. But I'm also hoping that he won't be placed into the direct action. Like you, I hope he'll stay in his van and direct the operation. Again, with the gear the company is providing, there should be no surprises when either of you are working your areas."

"Okay," Butch finally said. "What if things go completely to shit and I find myself having to shoot someone? Especially if it's one of the kids? I'm not sure I could shoot a little girl."

"I fully expect that part will be non-violent," Jim told him. "We'll have several ladies who work with Black Water doing most of that, regarding the massage parlors. I'm sure it will be less threatening for the girls to see a woman instead of a man trying to abduct her.

"And we certainly don't want anyone shot who isn't trying to harm you or any of our people," Jim added. "The folks who are truly on the ground for this operation are well qualified with firearms, and this isn't their first rodeo. Most of them have been on much more dangerous missions."

"How much luggage do I need to bring?" Butch asked.

"I'm bringing two pairs of socks, underwear, and knit shirts," Jim answered. "You might toss in an extra of everything since you'll be gone from home for one more day than I will. And please bring your toothbrush and deodorant.

"If you do need to get anything while you're there, Indy probably has a Walmart close by," Jim told him. "Don't get too wrapped up in the logistics. The company will make sure you're taken care of."

"Okay," Butch finally said. "I guess I'm so used to knowing where I'm going, what hotels I'll be staying in, and how long I'll be gone, this is all new to me."

"While you're thinking about things you'll need, let me know if you have any preference of weapons," Jim told him. "I don't know what your experience with firearms is, so it would help if you let me know what you'd be comfortable with."

"I used a .38 in the Air Force," Butch answered. "That's probably all I'll need."

"Why don't you discuss it with Mike?" Jim advised. "He's familiar with about every weapon on the market."

"That's a good idea," Butch agreed. "Well, I guess I'll let you go, and I'll call Mike and let him know what you've told me. Not to say anything out of line, but I get the impression that Mike is going to be happy being assigned to the TDA. I know he'll be happy he's not putting little girls on a bus. He's definitely more aggressive and more at home with more violent action than me. He's told me about wanting to go bear hunting with just a spear. Says he misses the adrenaline rush he got in combat."

"Well, I hope he doesn't get too aggressive here," Jim said. "This is a deportation issue. Not an annihilation. Anyway, I'll see you both tomorrow evening."

Next, Jim called Mike and discussed much the same as he had with Butch. After covering the details, he asked Mike what weapons he wanted when they got to Indy.

"A Kel-Tec KSG Bullpup twelve-gauge with four extra magazines and a Glock G18 with four thirty-three-round magazines," Mike answered.

"Sounds like you've given this some thought," Jim said, laughing. "Are you sure you don't need an M134 minigun like they used in that movie Predator? What'd they call that gun?"

"*Old Painless.* Too heavy. I've gotten fat and lazy in my old age," Mike replied. "If I can't handle these folks with what I've asked for, you may need to get a younger man."

"I'll put in your request. With luck, you won't need to use anything," Jim told him. "Sorry to disappoint you. And no, you don't get to keep the weapons after the mission is over."

"That's a shame," Mike replied. "I could sure use them with the hog problem up here."

"Don't forget those nasty chupacabra that have migrated," Jim said, laughing. "I'll see you tomorrow evening."

After hanging up, Jim thought about trying to reach Marie one last time, but decided she would know he had called, and she could call him if she wanted to.

Chapter 14

Jim was on his second cup of coffee when the phone rang. Again, wondering if it was Marie, he tentatively said, "Hello."

"Good morning, Jim," Gene told him. "If you'll look outside your front door, there's a package that has most of the details for both operations. It's the best we have for now and may need updating before the operation actually kicks off, but it's a start.

"And, since you have North and Knox coming over this evening, it will give you a chance to show them the scope of their areas," Gene added. "I've also included a list of the females who will be there to help with the Triad problem involving the little girls."

"Are there any Chinese speakers in the group?" Jim asked.

"Yes, but most of the people we're after speak English," Gene answered. "Not to mention, there are so many dialects of the Chinese language, we can't cover them all. The prominent language is Mandarin, and there will be at least one of our people who is fluent. That should be sufficient in the event some of the newly arrived little girls don't understand English. It's also sort of a guard against one of the older ladies giving them bad advice."

"Sounds solid to me," Jim said, opening the front door and getting the large manila envelope that was resting against the door.

"I talked to Butch and Mike last night," Jim said tossing the envelope on the table. "Butch is a little concerned with having to shoot someone and I believe Mike will look forward to it."

"That's pretty much the way you figured it would be, isn't it?" Gene asked.

"Yes, it is," Jim replied, retrieving his coffee from the kitchen. "But I believe he'll follow orders and with luck, he'll never leave the van he's using for a command post."

"I hope you're right," Gene said. "We sure don't need hundreds of videos or pictures of bloody bodies being drug out of the apartments or wherever we find the targets. It's going to be a political mess under the best of circumstances. Seeing dozens of bodies dripping blood on the sidewalk as they're carried out won't do much good for convincing the public how humanitarian this operation is."

"I know," Jim answered. "It would be nice if we could just run in, get the TDA folks, and escort them to the waiting vans or buses."

"That would be nice," Gene agreed. "It would also be a miracle of the highest level. We're just hoping that a sudden show of force will convince them not to resist. Just showing them being escorted in handcuffs from wherever we catch them will be enough to create a media firestorm."

"Understood," Jim said, opening the envelope. "What if I give you a call in an hour or two after I have a chance to review this new information?"

"Sounds good," Gene answered. "I'll talk to you then."

Jim spent the next couple of hours studying the locations and estimates of the number of people involved in the Triad operation. Seeing that there were seven separate strip malls where their operations were ongoing, and two apartment buildings where the girls were housed, it appeared that teams of four for each apartment should be sufficient.

On the opposite side of the Triad were the farms, situated several miles from the city. To coordinate operations effectively, teams would need to be dispatched before sunrise. This timing was crucial to make sure that all activities could commence within minutes of one another.

Doing this much work and still having a single target loose that can alert any other targets would be disastrous. The only way this could be coordinated was with computer assistance from Quantico. Hitting every place just after dawn would be ideal, especially since the public would more likely be at home.

It would also hopefully mean that the teams hitting the massage parlors wouldn't be stumbling upon some very unsavory activity. It would be much easier to gather the kids if they were still in the apartments where they were kept.

There were twelve one-bedroom apartments that housed eight girls each. The photos showed four bunk beds in the bedrooms, one against each wall, for a total of eight girls per room. The single bathroom would have made life there uncomfortable, to say the least. The information also included an additional apartment in each of the buildings where the supervisor of the girls lived. Thinking that the little girls wouldn't have phones, it would be impossible to locate them if they weren't in the rooms or the massage parlors. He'd have to focus on the ladies who controlled them. Catch them, and they could provide the location for any of the girls who are not in their rooms.

Jim sat looking at the names of the people who had been assigned to his areas, and he spotted one very familiar name, Jewell Ford. '*Damn*,' he thought, '*She finally got a field assignment. It'll be good to see her again.*'

An hour later, he called Gene to discuss the people he would be working with and pass along Butch's requests along with the weapons Mike had requested.

After Gene answered, Jim said, "I have a couple of additional requests, sir."

"What exactly are they?" Gene asked.

"First, the weapons for North and Knox," Jim answered. "Mike wants a Glock G18 with four thirty-three-round magazines and a Kel-Tec GS Bullpup with four extra magazines. Butch wants a .38 pistol."

"Is that all?" Gene asked.

"No, I want a car and driver, the same weapons as Mike, for both me and the driver," Jim told him. "And please make sure the driver you get is qualified with both."

"I don't think you will really need a car, driver, or the weapons," Gene replied. "I certainly don't want you out in the middle of this. With the number of people assigned and the scope of the operation, you need to be where you can direct the action from the safety of your command post. I don't want either Butch or Mike left to make the number of decisions that will be necessary to manage the entire operation."

"I hope you're right," Jim countered. "But how many times have things not been as planned? I want the capability to get to any of the locations if anything happens to either of them. This is their first operation, and I want the ability to get wherever I'm needed if something happens to either of them and I have to take over.

"And I'm betting that the little girls, or boys, don't have phones for Debbie to use to locate them. Just as with that little boy we found on Barco Do Amor, her intelligence isn't infallible. And there's always the not-so-impossible fact that a lot of the TDA gang changed their phones after the fiasco in Colorado," Jim continued. "I have enormous faith in her ability, but with so many people we're trying to corral, I'd be more confident if I could be more mobile.

"And, since I'll have constant connection with her and her team back in Quantico, as well as the glasses and iPad to help, I don't think it's an unreasonable request," Jim finished.

After pausing for a few seconds, Gene said, "Okay. You'll get your car and driver. And the requested weapons for both of you. Anything else?"

"Nothing I can think of for now that I'll need," Jim answered. "What can you tell me about the ladies who will be picking up the girls at the apartments, or wherever they are found? Specifically, Jewell. Is this her first time back in the field since her recovery time on the desk?"

"Yes, this is her first time," Gene replied. "Please don't tell me your little spat with Marie is making you think of starting something with her."

"No," Jim answered. "I'm not giving up with Marie until she decides whatever she does. I'm more interested in making sure Jewell is ready to get back into the field. A couple of near misses can make a person gun-shy, if you know what I mean. I'm thinking of the operation, nothing more."

"I made the decision to put her on your team," Gene answered. "Our psychologists say she's ready, and that's good enough for me."

"Good enough for me too," Jim said. "I'll let you know if I have any other thoughts when I see you in Indy. Have a good evening."

Chapter 15

It was almost four o'clock when Jim heard the doorbell ring. Opening it, he saw Butch and Mike standing there. "Come on in," he told them, holding the door open. "I didn't expect you guys for another couple of hours and definitely not together."

"I got restless and decided to grab Butch instead of taking two cars," Mike said as he looked around.

"We talked about the stuff as far as the Triad went," Butch told him following Jim into the kitchen. "I didn't know anything about the TDA folks or what was planned.

"We can talk about that right now," Jim said, taking three glasses from the cabinet. "I'm betting that you guys are ready for a little afternoon beverage. Same as at Butch's?"

"Sounds good to me," Butch said. "I may need an afternoon beverage every morning also until this is over."

"Maybe I'll join you," Mike said, laughing as Jim took a bottle of Jack Daniel's from a set of oak cabinet doors. Seeing the stock of liquor, Mike asked, "Do you always keep that much liquor around?"

"Pretty much," Jim answered, carrying the bottle to where he had left the glasses. "A man just never knows who might drop by for an afternoon chat. I can't be unhospitable, you know."

"I'd say you can be hospitable to half the county," Mike replied as Jim handed him a glass.

"It's a small county," Jim said handing a glass to Butch.

Picking up his glass, he raised it and said, "Cheers, my friends. May all your days be merry and bright?"

"Little early for a Christmas carol, isn't it?" Butch asked raising his glass.

"What the hell are you two talking about?" Mike replied, raising his glass.

"The song, White Christmas," Jim answered. "And the best version I've ever heard was by the Drifters. And the cartoon with the reindeer and Santa singing is my favorite."

"I know the one you're talking about," Butch said as Jim led them onto the back porch. "I send that one to my kids every year."

"You guys have a seat, and I'll bring in the latest info on the party in Indy," Jim said, setting his glass down. "This will probably be more to bring Mike up to speed, but it will let both of you know the scope of the entire operation."

Coming back to the porch, Jim pulled a couple of sheets of paper from the envelope and laid them on the table. "Not sure if you guys saw the fiasco up in Colorado," Jim said as he looked at one of the sheets. "This was the initial plan based on the TDA numbers before that happened. After the raid caused every TDA member in Colorado to leave, I'm sure there will be some adjustments."

Handing the sheet to Mike, he continued, "Gene has already told me approximately forty or so are expected to be joining their gang in Indy. And he is going to shift half of our people who were working there to join us.

"That should be sufficient if the TDA guys don't get too spread out," Jim explained. "As you can tell from the locations listed on the sheet, there are three apartments where they reside. We don't know if there is room for the guys coming from Colorado, or if they have another apartment or somewhere else for them."

"Will the people from Colorado, the ones you said were being sent up to Indy, be my responsibility?" Mike asked handing the sheet to Butch.

"Yes," Jim answered. "If the TDA guys go into the same apartments, it just gives you a few more for backup. If they wind up in another area, I'd suggest you use the guys from Colorado to cover that since you'll have already assigned your guys to the apartments."

"There are three apartment buildings," Mike offered. "How many apartments in each?"

"I think one of them, the closest to the airport, has about thirty," Jim said. "There's another a mile or so away that has about the same, and the last one only has fifteen. Most of them are one-bedroom apartments and have only been used by the TDA for less than a month.

"These places are more along the line of Section eight housing, and the TDA guys pretty much just took over," Jim explained. "There may be a few previous residents living there, but our information shows they ran everyone off."

"How many people are we talking about?" Mike asked. "And how many people will I have working for me?"

"Initially, there were about one hundred," Jim answered. "Toss in another forty or fifty if all the Colorado members show up as we expect. You are going to have fifty men to cover the three apartments."

"Fifty men to cover seventy-five apartments?" Mike asked. "That seems a little short to me. That's not even one per apartment. I'd like a minimum of three men for each one."

"First, just because there are seventy-five apartments doesn't mean each one is occupied," Jim told him. "These guys are used to four or five guys to the apartment, based on what we've seen. We'll get better information when we get to Indy, but I'm guessing they don't occupy more than twenty to twenty-five of the apartments.

"If the guys from Colorado join them, and we expect some contact and cooperation, there should be enough rooms in the three complexes," Jim continued. "And you'll get the twenty-five or so more of our people joining you. That's about seventy-five working with you to cover maybe thirty apartments.

"These places have ten apartments on each floor, five on each side of a common hallway," Jim said, handing Mike another sheet. "That'll mean about three men per apartment, pretty much what you want.

"Why don't you take this information and give me your plan?" Jim said. "If you don't think you can handle it with the men you're going to have, let me know and I'll see if the company can't find a few more bodies to help."

"Sounds fair to me," Mike said, nodding. "Sort of like back in Vietnam, send twenty eighteen-year-olds out to take on twenty nineteen-year-olds. Or vice versa."

"Let's just hope this ends better than that did," Jim said, picking up his glass. "Why don't we run grab something to eat and then hit the sack. Our plane will be at Love Field at noon tomorrow.”

“Are we flying Southwest?” Butch asked, following Jim into the kitchen.

“Private jet,” Jim answered, setting his glass in the sink.

“We must be moving up in society,” Mike joked, setting his glass beside Jim’s.

“Last trip I made was on the Greyhound after landing in the States.”

“You know what they say about riding Greyhound, don’t you?” Butch asked.

“What’s that?” Mike asked as Butch place his glass in the sink.

“Well, if you have sex in the airlines, it’s called the mile-high club,” Butch said following Jim to the door. “If you’re on the Greyhound, it’s called the thirty-six-inch-high club.”

Chapter 16

Jim drove them to Denny's the next morning for breakfast before heading to Love Field for the flight to Indianapolis. After ordering, Butch asked, "What information do we have about the farms? I pretty much understand the operation involving the parlors and apartments where the kids are, but what about those on the marijuana farms?"

"Best I can tell you is that there are five farms in our assigned area," Jim answered. "The company has already done the planning, which involves buses for the kids who reside on the farms, and any other people who are there when the operation begins.

"Once the kids are on the buses and the buildings cleared, the teams will take flame throwers and destroy the hothouses where the marijuana is being grown," he continued. "The rest of the farms are pretty much corn, wheat, or some other cover crop. The only things we're interested in are the kids and other illegal aliens. The illicit weed production isn't part of the deportation effort, but more or less a signal to the Triad.

"There has been some discussion regarding the foreign ownership of the land, but that's not our concern either," Jim told them. "And since much of that part of the Triad problem has been covered by the company, your involvement is minimal. Your main function will be to make sure the actual field agents have buses or solve any equipment issues.

"As with any of these operations, whether TDA or Triad, our responsibility ends when we deliver the illegals to the holding facility for whichever governmental group is given responsibility for their futures," Jim concluded as their meals were arriving. "Maybe there will be additional information when we get to the airplane or when we arrive at Indy."

A couple of hours later, as they waited in the lounge at Love Field, Jim saw a Gulfstream V pulling onto the ramp. Standing, he said, "Guys, looks like our plane is here. If you need to use the restroom, now would be a good time."

Mike stood looking out at the plane and asked, "Are you telling me that there isn't a head on that plane?"

"Of course there is," Jim answered. "But to be considerate to all of us, and maybe some other passengers, please take care of your business before we leave."

A couple of minutes later, Jim watched as the pilot with four stripes on the epaulets of his white shirt came down the stairs and headed for the General Aviation terminal, where they were waiting.

Walking in, the pilot looked around, then walked up to Jim and asked, "Mr. Lashley?"

"That would be me," Jim answered. "And these two guys are Mr. North and Mr. Knox. I'm sure you were expecting them and possibly other passengers."

"Just you three," he answered. "I'm Rob Sproc and I'll be your Captain for the flight to Indy. We don't need fuel, so if you'll follow me, we'll get on board and head out. By the way, there's a package for you back in the passenger area. I left it on one of the seats."

"Great," Jim said, grabbing his suitcase. "Grab your bags, guys, and let's get this rodeo underway."

"Everything ready back there?" Rob asked as the co-pilot closed the stairs.

"Good back here," Jim said looking at Mike and Butch sitting across the aisle from him.

Jim opened the package as the engines started and started looking at the updates to the information he had received from Gene previously. Looking specifically at the new data on the TDA numbers, he finally passed it to Mike, saying, "Looks like the guys from Colorado are holed up in some small ten-unit motel popular back in the fifties. I think this resolves the issue regarding the coverage of the TDA guys at the apartments. Just give this part of the operation to the Colorado folks and make sure we have vans and buses to support them.

"I'll check with Gene when we get there to make sure they have the other equipment, glasses, iPads, and stuff," Jim continued. "And I'll make sure Quantico gets you their displays and their phones are linked with yours. I don't want different organizations to work on the same problem. They'll be incorporated with ours, or they'll sit on their asses at the hotel until this is over."

"Are we all staying at the same hotel?" Butch asked.

"I don't know," Jim answered. "Normally, we do so that we're together, and that makes it easier to ensure everyone is fully informed every minute. But with this spread over such a wide area and such diverse groups, I can't say.

"I'd expect the guys going after the farms to be stationed close to where they will be operating," he continued. "And it would be reasonable to assume most of the actual field agents would be located near where they need to be tomorrow morning.

"I'm sure we will be with the folks who are responsible for the logistics and administrative crap are located," Jim concluded. "You'll be taking whatever equipment, basically your communications van, where you'll head for the most logical location based on where the targets are.

"I'll probably be stuck back in the hotel, probably some conference room the company reserved, and listening to you guys, listening to your guys, who are listening to each other," Jim told them as they heard the engines spool up for take-off. "At least that's what I'm hoping for. Listening to people do their jobs while we sit with a cup of coffee and a donut. But then I dream of winning the lottery also."

"What do you really think happened back in Colorado?" Mike asked as the plane started a gradual left turn to head north. "I mean, how did all of those TDA guys know they were about to be a raid?"

"I'd say that some Deputy Dufus told his wife to try to impress her how important he was. Then his wife's Sancho told another Sancho, who had some ties with the TDA, that a raid was planned," Jim answered.

"Who the hell, or what the hell, is a Sancho?" Butch asked.

Mike laughed and explained, "A Sancho is the Mexican term for the man who's sleeping with your wife."

Butch looked at Jim quizzically and asked, "Is that true?"

"Definitely," Jim said, laughing and handing him a folder. "But before you need to worry about Sancho, you must be married or seriously involved with a lady. And since you're neither, I think you should worry more about this new information on those girls who are taking care of the guys whose wives are being taken care of by their Sanchos."

"Ahhh, the circle of love," Mike said, shaking his head. "Guess it's what makes the world go round."

Chapter 17

After landing at Indianapolis International Airport and taxiing to the General Aviation Terminal, Jim watched as a black Suburban drove over to where they were parking.

"Well, gentlemen," Rob told them as the engines wound down. I'll have the stairs lowered as soon as the groundman lets me know the wheels are chocked and the area is clear. Thanks for flying with us."

"Thanks for the great flight, Captain Sproc," Jim said, getting the material from the package and his suitcase. "For some reason, I think I've flown with you before."

"Could be," Rob answered. "I've been flying for the company for several years."

"That's probably it," Jim replied as Mike and Butch got their bags. "Anyway, it's been a pleasure."

As soon as they got off, the driver met them and opened the rear of the Suburban for their luggage and said, "Good afternoon. I'll be taking you to the hotel where the General is waiting. Just sit your bags at the rear and I'll load them."

"We can toss them in ourselves," Jim told him. "Hope the traffic isn't too bad."

"Shouldn't be, Mr. Lashley," the driver said. "By the way, my name's Fabio and I'll also be your driver while you're here. I'm staying at the Conrad Indianapolis Hotel with you until this operation is over."

"Good to meet you, Fabio," Jim said, shaking his hand. "And please call me Jim. These two bewildered gentlemen are Butch and Mike."

"I know," Fabio said, shaking their hands. "I was provided a portfolio on each of you. Welcome to Indianapolis."

Jim got into the front passenger seat and asked as they headed for the gate, leaving the terminal, "How much do you know about the operation?"

"Probably just only as much as the General thought I'd need to know," Fabio replied as they headed for the exit across Loop 465 onto Sam Jones Expressway heading east.

"Who is this General?" Mike asked, leaning forward.

"General Gene Barker, USMC, retired," Jim answered. "He was in charge of the Marine Aviation in Vietnam. He's the guy who got me into MARCAD, the program where enlisted grunts like me got a job flying F4s, and ultimately American Airlines."

"Hell, I wish I'd known about that back then," Mike replied. "Maybe I'd have been a pilot like you and Butch. Free to chase all the Flight Attendants and drink all night."

Butch laughed and said, "Mike, I guess you know that the new guys have to pick from the sweet boy Flight Attendants."

"I doubt that," Mike said, shaking his head. "I can't imagine either of you even looking at them."

"New contract, new rules. Only applies to the guys hired after the contract took effect," Jim said, laughing as Fabio took the exit for US 70. "The boys demanded that they should have the same opportunity to marry a pilot as the girls. Our union agreed as long as we got a fifty percent pay raise."

"How do you think I managed to buy my ranch," Butch added laughing. "But if you pick the right boy, you can double your wardrobe."

"Don't forget the benefit of free flights," Jim told them. "That includes the opportunity to go on the trips with them to all the exotic places we fly. And the hotel room is free for the two of you."

"Assholes," Mike told them sitting back. "I'm not sure I want to ever be involved with you two ever again! Fabio, turn this damn thing around and take me back to the airport!"

"Sorry, sir," Fabio replied, grinning. "I've been ordered to deliver the three of you to the hotel. I'm just following orders, sir."

"You're both still assholes," Mike said crossing his arms and staring out of the side window. "Maybe all three of you."

"Not to change the subject," Jim said, turning back to the front. "Did the General tell you what being my driver could possibly entail?"

"More or less," Fabio answered as they approached Kentucky Avenue. "He asked me if I was familiar with the Bullpup 12 gauge and the Glock G18. I assumed that meant that I might be more than just driving."

"Possibly," Jim said. "Former military, I assume."

"Yes, sir," Fabio answered, glancing in the rearview mirror at Mike. "Marines, but I never wanted to be a pilot after I heard what the airlines required of the new guys."

"Fucking assholes, all three of you," Mike said. "I can't believe a brother Marine would show me such disrespect. Maybe from the Coast Guard. Hell, this is what I expect from an Air Force puke. But not a Marine."

"Semper fi," Jim said, laughing as Fabio turned north on South West Street.

"Sounds like the four of us ought to get together for a drink after this is over," Fabio said as they turned east on West Maryland Street. "Maybe we need to take care of your drink tab, Sergeant Knox. And by the way, the contract only applies to Navy pilots who come to American. They are used to hanging around the sailor boys."

"That's true," Butch added. "After all, one of the Village People was a Navy guy. They fit right in."

"Damn right you three owe me," Mike responded shaking his head. "And I have a feeling that it's going to be a long night tomorrow night."

"You probably don't know this, Mike," Jim turned to tell him. "The company foots the bill for all the food and drinks at the end of any operation. So, the drinks are on the General."

Turning north on South Illinois Street, Fabio said, "The Hotel is just ahead on the right after the next stoplight. I'll help with the bags, and you guys can go check in."

"We'll take the bags," Jim said as they sped through the light and pulled into the hotel. "I'm guessing you're staying here, also."

"Yes, sir," he answered, parking in front of the hotel. "The General is in room 5001, and he asked for you to give him a call as soon as you get in. I believe most of the group, including me, is on that floor."

"Sounds about normal for one of these things," Jim said as they got out of the car. "Guess I'll see you for this evening's briefing."

"Yes, sir," Fabio answered as Jim and the others headed for the entrance. "I'll just park the car, and if you need anything, I'm in room 5020."

"Thanks," Jim said, turning to watch Fabio shut the last car door. "Maybe we'll have a chance to get more into the driver's part of this little adventure."

Chapter 18

After checking in and taking his suitcase to his room, Jim called Gene's room and waited for him to answer. Looking out of the window, he realized that this hotel was probably one of the tallest buildings in the city.

"General," Jim said after the phone was answered. "Fabio said you wanted to talk to me."

"How was the flight?" Gene asked.

"Normal," Jim answered, picking up the sheets that had been in the package he got on the plane. "Does this have anything to do with the package you left for me?"

"Somewhat," Gene replied. "How about meeting me downstairs in the conference room? We're getting things set up for tomorrow morning, and I want to see what you think before the briefing later this evening."

"On my way," Jim said, putting the papers in his suitcase and setting it in the closet.

A few minutes later, Jim walked into one of the largest conference rooms for a hotel he had ever seen. At the back of the room was a screen that covered most of the wall, and at least forty desks were set facing the screen. The podium faced the door, and approximately one hundred chairs were arranged in rows between the door and the podium. Tables lined the walls on three sides.

"What do you think?" Gene asked as he walked in and stood beside Jim.

"Impressive," Jim admitted. "I've seen briefing rooms half this size at the Pentagon. What are the tables along the walls for?"

"The hotel will be providing a buffet for this evening's briefing and again for breakfast tomorrow morning," Gene answered. "Then they will clear the tables and provide another for lunch."

Walking toward the podium, Jim asked, "Are the Colorado guys going to be here for the briefing?"

"Just their team leader," Gene answered. "Do you want to meet with him before that?"

"I'd like for Mike and me to meet him before the briefing," Jim answered, turning to look at Gene. "I want to make sure there are no turf wars here. Mike needs to be confident that these folks understand that they are ultimately under his control, and if he needs to shift some of their people, it's his call."

"How soon can you have Mike down here?" Gene asked, pulling out his phone.

"Ten minutes," Jim answered.

"Call him," Gene said as he waited for his phone to be answered.

"Jon," Gene said, turning away from Jim. "How soon can you be here? There are a couple of people I'd like you to meet before the briefing."

Hearing the answer, he disconnected and said, "Twenty minutes. Tell Mike to be here then.

"Let me run over a couple of things while we're waiting," Gene said, heading to the area where the desks were facing the screen. "You were correct about the TDAs from Colorado. They ditched their phones.

"That's not going to be a problem," Gene continued. "Debbie's voice recognition program found their replacements shortly after they got here, with a couple of exceptions."

"And those are?" Jim asked.

"Three of the guys didn't get new phones. At least we can't identify any phone they are using," Gene admitted.

"So, we're going to have at least three unknowns at the motel where they are staying," Jim replied. "Or even three unknowns who could be just about anywhere, including the apartments."

"That's the situation," Gene admitted. "We're still hoping Debbie can find them and get their phones into the system."

"Anything else?" Jim asked as he thought about how the three unknown people could possibly impact the safety of his people.

"A little, more on the good side of the ledger," Gene told him. "All of the phones, including those we've provided the Colorado guys, have been modified with an encoding program.

"Granted, it's a very simple program," Gene continued. "It could be cracked by a half-witted hacker within a couple of hours, if he had enough transmissions. But I don't think we'll be dealing with anyone with that skill set. So, I'm confident that any conversations will be secure regarding all of the operations."

"I might agree with you regarding the TDA," Jim countered. "But the Triad has some pretty sophisticated networks. I'm not sure they can't get around any simple code within a few minutes. Especially if they're getting information about raids from across the country. Their command-and-control system must be much more centralized than the TDA."

"You're right," Gene told him. "But each of our operations across the country will be using a different encryption. If they do break any one code, they still won't know what's happening in the other cities."

"That's a small comfort for me," Jim said, shaking his head. "I don't want to be the city where they break the code."

"What do you suggest?" Gene asked as they saw Mike coming into the room.

"We've got almost twelve hours," Jim answered. "Let's see if Debbie can come up with a system such as we had before. Remember the old 'Have Quick' radios? Maybe she can program all the phones here in Indy to frequency hop every second or so. It doesn't have to be as sophisticated as the Have Quick, but technology has outdated that system anyway."

"I'll give her a call right now," Gene said. "If she can do that for the phones here, she should be able to take care of all the phones we're using across the country."

As Gene turned and walked away, making his call, Mike walked up to Jim and said, "This is a little more than I had expected. I'm used to briefings of ten or fifteen folks in a tent with folding chairs. This company must have some major financial access."

"I'd say so," Jim replied. "Gene asked the head of the Colorado boys to be here shortly so we can make sure we're all on the same page. He should be here in the next few minutes. There's a small hiccup in one of the things I thought we had covered, but I'll wait until the Colorado guy gets here and explain it."

"How small is this hiccup?" Mike asked, walking toward the screen, looking from side to side.

"There are potentially three Colorado TDA guys we've lost track of," Jim answered as Gene turned and came toward them.

"Three out of the fifty or so coming from there shouldn't be a problem," Mike said as Gene stopped beside them. "I've faced worse."

"We don't want you facing worse here, Sergeant Knox," Gene said, extending his hand. "I'm Gene Barker, and I'm responsible for this little operation. And that means I'm responsible for ensuring your safety."

"General," Mike said, coming to attention as he debated saluting or taking the extended hand. "I'm glad to meet you."

"Likewise," Gene said, reaching over to take Mike's hand. "And it's just Gene now. Jim has told me a little about you, and I've done a little research. Two Bronze Stars with a 'V' for valor tell their own stories. Glad to have you here."

"Thank you, sir," Mike said. "I hope I can live up to your expectations."

"I have no doubt that you will," Gene told him as another man entered the room. "Just as I have no doubt that any of the people we've selected will take care of the issues we're facing.

"Jon, thanks for coming," Gene said as the man joined them. "Let me introduce you to a couple of the guys you'll be working with. First, Jim Lashley. He's been with the company for several years. Worked with Dark Water the first few years and transferred to Muddy Water after retiring from the Marines.

Waiting for them to shake hands, he continued, "And this is Mike Knox. You'll be working directly with him, and he'll be reporting back to Jim."

Shaking Jim's hand, Jon said, "Good to meet both of you. I was sort of surprised when the company sent us up here after all the planning we had done earlier."

"That was such a screw-up," Jim replied. "We're trying to make sure the same thing doesn't happen here."

"Speaking of that," Gene told them, "I've gotten assurance from the company that all of the agencies involved in Colorado's disaster will be remaining clear of the cities where we have operations."

"How did they arrange that?" Mike asked.

"You obviously don't know how far the reach of this company is," Jim told him. "I'd be willing to bet that someone at or near the top of the political food chain was told that any interruption in our operation, or any leak about it, would result in some very negative publicity. And most likely some, let's call it shunning, the next time these folks need our assistance."

As Gene's phone alerted him of an incoming call, he turned away and listened for a few seconds, and then announced, "Looks like you got your wish, Jim. Bracer says she can have all our phones synced to change frequencies or numbers every thousandth of a second. That should resolve the potential problem with the Triad."

"But we still have the problem with the three unknown TDA guys," Jim said. "And that's what will impact these guys. Not to mention, I'd rather have three short, chubby Chinese madams from the massage parlors running loose than three guys who are willing to shoot me in the face."

Chapter 19

A few minutes later, a lady walked in and stood quietly beside Gene until he finally said, "Jim, this is Amanda. She's your communications assistant. If there are any issues or requests, she's your lady."

"Nice to meet you, Amanda," Jim said, taking her outstretched hand. "I wasn't aware I needed an assistant for any communications, but I never turn down someone who wants to help me. Just what will you be doing for me?"

"Basically, I'm here to monitor all communication coming in and ensure that any from the team leaders, and Mr. North and Mr. Knox, are routed to you and recorded if we need to review the conversations.

"I will also have a hotline with you, Mr. North, and Mr. Knox," she continued. "If the phones are becoming overloaded, as they sometimes do with an operation this large, I can prioritize the incoming calls so that you won't miss anything from those people."

"Sounds good," Jim said, nodding. "I agree with the number of folks we're working with; it could get a little confusing since we're hearing every phone call and conversation we're having."

"And don't forget the phones our targets have," Amanda reminded him. "I'll be able to monitor those and relay them to you if they have any potential information."

"That alone is worth having you here," Jim said turning to Butch and Mike. "By the way, this is your Mr. North and Mr. Knox. Guys, meet Amanda."

"Nice to meet you," Butch said shaking her hand. "And please call me Butch."

"Nice to meet you, too, Butch," she told him. "And please just call me Amanda."

"I'm Mike," Mike said interrupting the moment where Butch and Amanda were looking at each other. "And since Jim only introduced you as Amanda, guess that's what I'll call you, too."

Amanda turned from Butch and shook Mike's hand saying, "Mike, good to meet you."

"I heard what you said you could do for Jim," Mike told her. "Can you do that for me and Butch as well?"

"Definitely," she answered. "I have software in my equipment that recognizes voices, and I'll program you, Butch, and Jim, in as priority. If I determine there are any other people necessary to prioritize, I can add them somewhere on the program.

"For example, let's say we determine one of the targets is the leader," she explained. "I can put him just below you three and make sure we monitor every conversation he makes. That will also record his conversation, just as it will yours, in case we need to review it. I can pull his conversation, or any portion of it, from the data bank and replay it within a couple of seconds if I, or any of you, want to hear it again."

"How many people can you have on your priority list?" Mike asked.

"Unlimited," she answered. "But what good would it do to have number four thousand eight hundred and seventy-two? Just take up data space and retrieval time. If they are that far down the list, I doubt they have anything to say that's very important. If we thought so, they'd be higher on the list."

"Given that," Jim said. "I'm assuming that you can take any conversation and place it higher in the priority list. For example, we hear one of our targets is leaving his location to go somewhere else. Or that he's heard information that we need to know.

"You can then put all of his communication higher on the list," he finished.

"Definitely," she answered. "I can block any communication that seems to be irrelevant or personal, such as Butch calling his wife or girlfriend, and have it return after a pre-determined lull in the communication."

"If you hear me talking to my wife or girlfriend, you've got a problem with your voice recognition system," Butch said smiling at her.

Amanda gave a slight nod, glanced at Butch, and said, "If any of you would like a quick demonstration after the briefing, I'll be at my desk up by the big screen."

As Amanda was walking away, Gene brought another lady over and introduced her, saying, "Jim, this is Tammie. She'll be your contact for any equipment needs. Tammie, Jim."

"Good to meet you, Tammie," Jim said. "I guess you got all of my requests for weapons up to now."

Tammie shook Jim's hand and answered, "Let's see three 12-gauge Bullpups with three-inch double ought magnum shells, twelve extra boxes, three Glock 18's with the extended thirty-three round magazines, and a .38 revolver."

"And a car for my driver," Jim added with a smile. "You seem to be on top of things."

"If I'm not mistaken, your driver has already taken possession of the car," Tammie replied. "But I'm here for anything else you may require."

Jim turned to Jon and asked, "Is there anything you or your folks need?"

"Not really," Jon replied. "We brought all our stuff with us, but I'd love to have one of those Bullpups. If it isn't too much trouble."

"I'll have it for you after the briefing," Tammie told him. "Anything else?"

"Well, I do love Snickers," Jon answered, smiling. "Or Butterfingers."

"I've left you a box of both at the Chevron station on the corner," Tammie said, handing him a card with just her name on it. "All you have to do is hand the cashier my card and then give him enough money to cover the purchase."

"I guess you'll reimburse me later," Jon said, looking at the card.

"Of course not," Tammie answered. "But you can keep the card just to remember me."

Turning from Jon, she looked back at Jim and said, "If you find tomorrow that you, or your people, need anything, let me know. I have an open account at one of the local gun dealers and can have just about anything from additional weapons to ammo headed your way within minutes."

"That's good to know," he answered. "I'm hoping we have enough, but I'm glad to know there are supplies handy. The biggest problem is that we don't normally have minutes when we run out of ammo. And we don't usually know we're about of ammo until it's too late."

"I understand," Tammie replied. "But the option is there, if you see the possibility of an issue, I'm a call away."

Jim motioned for Jon and Mike to join him as he walked to the side of the room. As they stood around him, he asked, "Jon, do you see any difficulty working with Mike?"

"Absolutely not," Jon answered. "Just as I have no problem working with you. The guy we had in charge was sent to Chicago to help up there since many of our targets fled in that direction. I'm here to work with whoever the company says is in charge."

"Good," Jim said, nodding. "Now, did you get the information regarding the three TDA members whose locations are unknown?"

"I did get something along that line," Jon answered. "I've sent a couple of my men to stake out the motel where we were told they would be. I can't confirm it, but there are a couple of guys there we haven't got a good look at, but I suspect it's two of the three.

"And if it is, the third one is close by," he continued. "I watched these shitheads for over a week while we were planning our raid. Then the big boys decided they needed to see themselves on the news and our plans went to shit. But I got a pretty good look at most of those dirtballs. We'll have an ID on the two unknowns at the motel before the sun goes down. And probably number three as well."

"Did you guys bring night vision goggles?" Jim asked.

"No," Jon answered. "We didn't need them back in Colorado."

"Hang on a second," Jim told them, turning away.

Walking up to Tammie, he asked, "Do you have access to NVGs?"

"Night vision goggles?" Tammie responded. "Of course, how many do you need?"

"I'll send Jon over to let you know," Jim answered. "How many gun dealers are there in Indianapolis?"

"Hang on a second," Tammie told him, taking out her phone.

"I'll be back in a minute or two," Jim said, turning away.

Jim walked over to Gene and said, "Sir, may I make a request?"

"What do you need?" Gene asked.

"Flashbangs," Jim told him.

"How many are you talking about?" Gene asked.

"About one hundred," Jim answered.

"Where do you plan on using them?" Gene asked.

"In the apartments and motel where the TDA targets are located," Jim answered.

"Give me a couple of minutes," Gene answered, pulling out his phone. "What about the farms and where the girls are being kept?"

"No, I don't want to make things worse there," Jim told him. "Those kids are going to be traumatized enough with the troops storming in. Besides, I don't expect much resistance from them."

Gene held up one finger and asked into the phone, "Sir, need some assistance rather quickly."

Listening for a moment, he continued, "We need a hundred flashbang grenades for our operation here. Do you have a source that can have them here this evening?"

Looking at Jim while he waited for the answer, he finally said, "That works for me. Thank you, sir."

Hanging up, he told Jim, "There will be a hundred grenades here in about two hours. Is that all?"

"For now," Jim answered. "I think I have a handle on everything else."

"Let me know if I can help with anything," Gene told him.

"I'll do that, and thank you, sir," Jim told him, turning away.

"There are eleven gun shops in the area," Tammie told him as he walked up.

"Okay, I want you to go to each one and buy one set," Jim said. "If you get asked, just tell them it's for your husband's birthday since he is a big deer hunter."

"What if he asks for identification?" Tammie asked.

"Show it to him," Jim replied. "That is, unless you happen to have a spare with someone else's name."

Tammie reached into her back pocket, pulled out her wallet, and showed Jim a fake ID showing her as Becky Martin. "Like this?" she asked as Jim took the ID.

Turning it over, he noted that it had an Indianapolis address and her picture. "Always carry a fake ID?" he asked, handing it back.

"Only when working for the company," she answered, putting it back in her wallet.

"Do you need a driver?" Jim asked, shaking his head at hearing her answer.

"That would help," she answered.

Jim looked around and spotting Fabio said, "Hang on a second and I'll see if I can arrange that."

Walking over to Fabio, he asked, "Can you take care of something for me right quick?"

"What do you need?" Fabio answered.

"I need you to take Tammie to a few gun dealers here in the city and pick up some NVGs. It would be prudent to park where you can't be seen from the store so nobody can spot the car."

"Not a problem, sir," Fabio answered. "When do I need to leave?"

"Right now," Jim answered, leading him to Tammie.

"Tammie, this is Fabio. He's my driver but has agreed to take you to the stores. Fabio, this is Tammie. She's in charge of these things, so take good care of her," Jim said as they shook hands. "If anything important comes up during the briefing, I'll let you guys know."

"We'll be back as soon as we can," Fabio said. Turning to Tammie, he asked, "Do you have the addresses?"

"Here on my little navigator phone," she answered as he led her to the exit.

Spotting Mike talking to Jon, Jim walked over and said, "Excuse the interruption, but I have some quick updates for both of you."

"What's that?" Mike asked.

"I've sent Tammie for some NVGs," he answered. "To keep from drawing anyone's attention, she's getting one from each of the eleven gun dealers here in the city. I know you'd probably like more, but that's all I can get given the time we have. Also, I'm getting about one hundred flashbangs. Figured you could use them before busting into the motel or apartments. Might save some bullets flying around the rooms.

"I'm not trying to disrupt your plans, Mike," Jim said, looking at him. "I just thought you might like to have them at your disposal. Use them as you think best."

Mike thought for a minute, looked at Jon nodding, and said, "Great. I didn't think about that. Could come in handy. Thanks."

"No problem," Jim replied. "Just looking out for my men. Now, I'll let you two discuss any changes you might care to make."

Chapter 20

An hour later, Gene found Jim talking to Butch and asked, "Can we grab something to eat and have a seat somewhere?"

"Sure," Jim said turning from Butch and the few men standing with him. "I'll follow you."

Gene selected a few items from the buffet the hotel had finished setting up, took his tray to one of the chairs at the back of the room, and waited for Jim to join him.

"Anything special you want to discuss?" Jim asked, sitting in the chair to his left.

"I just wanted to bring you up to date on a couple of issues," Gene said, setting his glass of tea on the chair on his right. "First, the flashbangs. I can't promise they will be here in two hours as I first told you. I considered getting the company involved with the locals but dismissed that because we would be providing them with knowledge that something was happening.

"I looked at every option I could think of to make sure nobody knew what we were going to do," he continued. "When I realized that the only way to get the grenades and

110

keep it somewhat under wraps was to have the company source them from anywhere except where we were going to conduct raids tomorrow morning.

"I just received word that we won't have them until maybe an hour and a half before we start the operation," Gene continued. "I know it puts you in a pinch, but it was the best I could do without some dimwitted deputy from the local arsenal wondering why so many grenades were being requisitioned.

"The other option was to proceed without them," he finished. "Worst case, you're no worse off than before you decided you needed them."

Jim thought for a minute and then replied, "I understand. I'd rather do without them than have anyone local with any suspicion about this. How are the grenades coming in?"

"They are coming in on one of the company airplanes," Gene answered. "Damn sure couldn't put them on a commercial flight. They'll be landing at Indy International and parking at the same terminal where you were picked up today."

"Hang on a second," Jim said, setting his tray on the seat beside him. "I want Mike's opinion on this since it's his operation."

Walking over to where Mike was talking to Jon, he said, "Mike, hate to interrupt, but something has come up and I need you to come with me for a couple of minutes."

"No problem, Jim," Mike responded, turning to follow him. "What's this about?"

"The flashbangs," Jim answered, leading him to where Gene was waiting.

"Gene just told me that the flashbangs won't be here until a little over an hour before you're due to hit the apartments and the motel.

"If you want them, you need to come up with a way to get them from the airport to your teams spread across the city," Jim told him. "Any suggestions?"

Mike stood quietly for a minute and answered, "What if I meet the airplane and take the grenades to the teams?"

"I think that would take too long," Jim answered. "But maybe if you find a central location and have a member of each team meet you there, you could give them the grenades they need for their part, and they could take them to their location. That could save you at least half an hour."

"That seems doable," Mike told him, nodding. "What if I'm at the airport waiting in my van? I can still hear from my folks and be ready if the airplane's early."

"That's fine with me," Jim told him. "But, should things go to shit, and the airplane not make it in time to deliver the goods, I don't want you to delay the operation.

"This thing is coordinated with other teams across the country," Jim explained. "We don't know exactly how interconnected these groups are, but we can't afford to take the chance that one group learns of mass raids before we kick in their front door. The best outcome if that happened is they are gone when our folks bust in. Worst is that they are ready and blow the shit out of our people.

"And when you think about it, if you delay, you're the last ones in and the most likely for your targets to know you're coming," Jim finished. "Understand?"

"Definitely," Mike said, nodding. "It would be nice to have them, but my guys can take care of our side without them."

"Speaking of your guys, have you met the other team leaders?" Gene asked.

"Missing one, sir," Mike answered. "He hasn't gotten here yet, but I've been assured he'll be here for the briefing. I'll bring him up to date. Now, if you'll pardon me, I'm going to sit down somewhere and see if I can find a location that would be optimum for getting the grenades to my folks."

"Just a thought," Jim said. "You might consider making the apartments a priority. Since the motel is fairly easy to contain with only one door to each room, they would probably be able to contain the targets. The apartments usually have a secondary exit for emergencies, such as fire escapes."

"Good idea," Mike agreed, nodding. "Anything else?"

"Yeah, the NVGs," Jim answered. "They should be here before the briefing is over and you can give them to your teams as you see necessary. Not to tell you how to do your job, but since there are only eleven NVGs, I'd concentrate on the apartments for the same reason as the flashbangs. Two possible exits versus only one."

"Got it," Mike replied. "I'll get with all my teams and take a look at all the locations, and if there are any escape holes. By the way, can someone stay in contact with me regarding the arrival of the plane with my equipment?"

"Talk to Amanda," Jim advised. "She's in charge of communication and can probably put an icon on your iPad showing the location, track, and speed of the airplane."

"Sounds good," Mike said, looking around. "Have you seen her lately?"

Gene answered, "I think you'll find her over there at the buffet with Butch. They seem to have developed a certain, let's call it mutual admiration."

"Have Butch come over here when you get there," Jim told him. "I need to make sure we cover his part before the briefing starts."

"I don't suppose you're giving him any of the stuff you've promised me, are you?" Mike asked.

"No, we don't want to scare the little kids," Jim answered. "Besides, I'm sure Gene will agree with me when I say our number one priority is the TDA. Although child sex trafficking is abhorrent, a massage with a happy ending has never killed a man. Nor has there ever been a case of death by nail polish."

"No argument from me," Mike replied. "After reading the rap sheet on some of these TDA guys, I'm more inclined to just ship the bodies home. You can stack several boxes of bones in the same space as a row of passenger seats."

"Just remember the rules of engagement, Sergeant," Gene reminded him. "And remind your teams. We may not like the way the rules were written, but with all the cameras and bleeding hearts wandering the streets looking to become YouTube stars, the less blood on the streets, the better."

"Guess I'll tell my folks to shoot them in the privacy of their rooms," Mike replied, shaking his head. "No problem, sir. I'll stress your position and make doubly sure they understand."

"Thanks, Sergeant," Gene told him. "I have no doubt you'll do your best to keep this out of the news. And off YouTube."

"Yes, sir," Mike replied, coming to attention.

"Once a Marine, always a Marine," Gene said, chuckling. "But he's wound a little tight."

"Probably got his ass chewed out by a General or Colonel before he was discharged," Jim replied. "I've seen some former enlisted Marines damn near salute when they meet a former officer. Sort of like Pavlov's dog. Once conditioned, they can't help but respond. It may be fairly minor, like straightening their back or snapping their heels together. But it's there if you're paying attention."

Chapter 21

A few minutes later, Butch came to where Jim and Gene were finishing their meals and asked, "Did you need something, Jim?"

Jim placed his empty tray back on the seat he had vacated, stood, and answered, "Not really. I was just wondering how it was going with you. Have you met all your team leaders?"

"Yeah," Butch told him. "They all got here early, guess for the gourmet buffet, so we've had a chance to discuss the operation."

"Is there anything you need?" Jim asked. "I'm guessing that the farmers and parlor people have differing requirements."

"Yeah, there's quite a disparity there," Butch agreed. "But the teams going after the 'farmers', as you like to call them, have a pretty good handle on everything. They've spent some time watching their targets and have an excellent plan to get the kids and destroy the marijuana crops. And the hot houses."

"Any transportation issues?" Jim asked.

"Nope," Butch answered. "Each of the five farms has fifteen to twenty kids working there. There's also a man and a woman who live on the farm. Plus, what I've decided is a supervisor when the couple needs him. He also doubles as security in the event strangers approach the farm. I don't know if the couple or the security guy are illegal or not, so I've told them to bring them with the kids. Let ICE separate the wheat from the chaff.

"The average school bus will hold forty-two adults, or seventy-two kids about the size of an eight-year-old," Butch continued. "These folks are significantly smaller than their American counterparts, so there is plenty of room for everyone, including the three men and driver from each team that raided the farm. They will provide security on the bus until we deliver them to the Feds."

"That sounds good," Jim said. "What about the parlor people?"

"I like that," Butch replied grinning. "Parlor people and farmers. I like that. I've just been referring to them as the boys and girls.

"As to the parlor people, my folks have managed to get keys to all of the apartments where the parlor kids are being kept," Butch continued. "The two apartments aren't close enough to each other to make it a single target, so there are two buses. Each apartment building has six apartments for the kids and one for the supervisor, who I've found out is pretty much the madam at the massage parlors. Anyway, that's forty-eight girls and the sorry piece of humanity, who whore's out these little girls, plus the Mandarin speaker, and two guards per bus."

"How did you happen to get the keys to the apartment?" Jim asked.

"One of the team went to the apartment complex with his girlfriend, or wife, or something, under the pretense of looking for an apartment," Butch explained. "The girlfriend managed to get an impression of the keys on the keyring he left when he opened the door to show them an apartment above those where the kids are.

"To make things easy, the brass keys had the apartment number stamped on them," Butch continued. "Our guy simply took the impressions to a locksmith friend, and keys to the apartments appeared the next day."

"Unbelievable," Gene said, shaking his head. "I guess they went back to the apartments and made sure the keys fit."

"Definitely," Butch answered. "And while they were there, they managed to place a couple of small devices in each of the apartments to monitor them. Including the madam's apartment."

"Shit, I may as well go home," Jim said smiling. "Just give me a call when you return the buses."

"Well, there's still the issue of who is and who isn't an illegal, or here on a visa," Butch said. "I told my guys to treat the madam the same as the farm folks."

"Best approach I can think of," Jim told him. "There isn't any way for us to know if they are US citizens, have valid visas, or anything else. Let ICE take care of these ladies also.

"I have one other request," Jim said. "I want your guys to go to the parlors after you've rounded up everyone in the apartments, and make sure there aren't people there. It's possible that some kid was brought in recently and is being kept separate from the others. I don't want a single stone unturned."

"That shouldn't be a problem," Butch replied. "Once the buses leave, I'll send everyone not on the buses to the parlors where these kids worked. Do you want us to look at the nail salons and other stores owned by the Triad?"

"Definitely," Jim said. "We also suspect drug activity has been taking place there. When your guys get in, search every crevice and photograph every inch. On second thought, have everyone run by Walmart or some store and get several Ziplock baggies. Get quart and gallon-sized just in case you need to collect samples."

"What do you want us to do with them, if we find anything?" Butch asked. "The buses will already be gone."

"This is more a DEA area than immigration," Jim reminded him turning to Gene and asking, "What do you think, sir?"

"Definitely not related to immigration," Gene agreed. "Let me think on this and make a couple of calls. I think this gives us, the company, the opportunity to give the guys who hired us a little bonus. They can probably parlay it into some political points. Damn sure wouldn't hurt to put thousands of pounds of drugs on the morning news that had direct ties to the Triad's operations. I'll have your answer before morning."

"Anything else, Jim?" Butch asked looking over his shoulder to where Amanda was sitting.

Jim noticed him looking away and said, "Not unless you have anything you need from me."

"I'll let you know if I do," Butch answered. "I'll have my guys make whatever adjustments are necessary to make sure the raids on the parlors are as you wanted."

"Thanks, Butch," Jim said. "Again, let me know if you see any issues with that. Sorry to toss this at you at the last minute, but I'm sure you can handle it."

As Butch walked away, Gene said, "I think you picked the right guy for this part of the operation. I had my doubts earlier, but he seems a perfect fit. Seems to be very detail-oriented, and that's what I like to see."

"Yeah, he may not be the guy you want beside you in a running gun battle," Jim replied as he watched Amanda put her hand on his arm when he got back. "But he'll make a damn good planner. That is if the company decides they need him in the future."

"What do you think of him and Amanda?" Gene asked. "Will we have a problem with any harassment issues?"

"I don't think either of them are thinking about harassment complaints right now, do you, General?" Jim answered.

"Not unless dissatisfaction is harassment," Gene told him. "Sort of like the lady who didn't know she had been raped until the check bounced."

Chapter 22

After it appeared that everyone had eaten and was just standing around in groups, Gene walked to the podium and asked everyone to please take care of their trays and grab a seat. A few minutes later, when almost everyone except a few still putting their trays in the receptacle beside the buffet table had taken a seat, he turned on the screen behind him.

"Ladies and gentlemen," he began. "I have a few words to say before we get down to the serious business that brings all of us here today."

Waiting another couple of minutes, he continued, "As most, if not all, of you have heard about the disastrous ending to the first mass raid on the exemplary TDA visitors down in Colorado.

"We've gone to extraordinary effort to ensure that doesn't repeat itself with our operation," he explained. "We've kept most of the politically minded people from knowing when or where we're going to strike tomorrow morning.

"I'm not talking about just here in Indianapolis," he continued. "We're conducting similar raids across the

United States. Now, to keep word from spreading, the raids will begin simultaneously. Each operation will begin at 0630 Eastern time. That's 0530 in Texas, Central time, 0430 in Arizona, Mountain time, and 0330 in California, Pacific time.

"You should be thankful you're not in California," he said, smiling. "At least here, you can get a reasonable night's sleep."

Waiting for the murmur of the people to die down, he continued, "If you're wondering how the targets we've been assigned were chosen, it's fairly simple. Those cities that are designated as sanctuary cities were assigned to Black Water.

"And before you show how politically astute you are, I realize Indiana prohibits sanctuary cities, as do some other states. But what's the law, or policy, isn't always the way things are conducted in reality. The criteria used were more local, as in the city of Indianapolis, where the Mayor is Democratic and between him and the Chief of Police, their support for this operation was questionable.

"Some state/city combinations are downright inhospitable to our country's authorities who are trying to uphold US federal laws," he told them. "They, our elected government, decided to do an end run around their local law enforcement and use our more covert operation.

"Those state/city combinations that actively support the DEA, ICE, Homeland Security, Border Patrol, or whoever is trying to remove those who are here illegally, especially those guilty of crimes besides just illegal entry, will be handled by those organizations I've mentioned," Gene explained. "Their timetable is different than ours. It was agreed that we would be the first so as to minimize the efforts of those who disapprove of the overall effort to remove these visitors.

"Before you say anything about the fiasco in Colorado," Gene told them as the screen behind him showed the masses of police and other law enforcement organizations. "This was not supposed to happen. That location was originally in our contract. Whoever decided to jump in and get some news time did all of us a disfavor."

Pausing as the screen showed multiple pictures of the hundreds of law enforcement personnel, he continued, "Black Water assured the folks in charge of deportation that one more incident, and we would remove ourselves from the issue.

"Now, with that in mind, the reason all of you will be wearing the uniform of one of the aforementioned groups is to provide a face for the public to recognize," he explained. "If you happen to be photographed or videoed while conducting your assignment, that will appear to be one of the organizations that are tasked with the raids.

"Between the uniform, the logos on the vehicles, your helmet and face shield, you should remain anonymous," he told them. "If approached by either a civilian or law enforcement, just politely tell them to call 800-328-7448. That number will route them to Black Water operations, where they can talk to someone who authorized your operation and further direct them to whichever organization whose uniform you are wearing.

"We, Black Water, through our domestic side, Muddy Water, have spent enormous time and resources getting us to this point," he explained. "As far as we can determine, no one outside our organization has an inkling as to what will take place across America in approximately eight hours.

"As you know, your phones have been set up to send your precise location back to Quantico, where hundreds of our most capable people are monitoring you and any target in your location," he assured them. "In a moment, I'm going to ask each of you to make sure your phones are on and on speaker.

"Our computer wizard, Debbie, will run a final check to verify its proper operation," he told them. "She will also activate a new program she's installed, while you slept, which changes the frequencies, or more correctly, the phone numbers of all the phones, to make sure no one can listen to any phone number and gain any information regarding this operation.

"So, if you'll check your phones, I'll get Debbie on the line," Gene told them as he dialed her number from his phone.

"Debbie," he said as she answered. "If you'll hang on a second, I'm going to transfer control of the screen here where we're conducting our final briefing before heading out to our locations, to your control."

As the screen suddenly showed the layout of the conference room, Debbie said, "Good evening, everyone. If you'll watch the screen, I'm going to jump around the room to designate several people. When that person's phone rings, I'm asking him or her to stand up and remain up until I finish."

Suddenly, a green icon on the screen flashed and Jim shook his head as his phone rang, and he rose from his seat. Next came Gene's phone. This continued until half a dozen people were standing.

"Now, I'm going to activate the random phone number program," she told them. "You will not notice anything different except the screen will begin showing the location of the person who was assigned the current number. Within half a second, the screen will show the next number in sequence."

As the screen began flashing from one location in the room to another, she explained, "Every phone here is on this system. You won't know it because it's changing frequencies, numbers so fast that unless you speak faster than a speeding bullet, your conversations will be perfectly coherent. No breaks, skipping words, or interruption of the normal use of the phone. Suppose you are calling Jim, for example. His phone will ring even though the number or frequency you'll be talking on will be changing. The only thing that affects the number generator is if a phone's original number is being used.

"Using the example of Jim's phone being in use, I'll demonstrate," Debbie said as Jim's phone rang. "Even though the number of his phone was in the random generator, once that phone is in actual use, that number is removed. The only way this system can fail is if everyone's phone is being used simultaneously.

"Now, even though the numbers are constantly changing, our computers identify each phone by an ID number," she told them. "That ID number will remain with the phone so we can use it for location purposes. You've already had this happen when you trade in your old obsolete phone, which seems to happen every five weeks, at your service provider. The old phone's number is transferred to the new phone. The old phone is sold at some pawnshop, and it gets a new number from the new owner's service provider. The old phone can still be tracked by its ID, even though it has a new number."

As the screen returned to constant green icons as originally seen when Debbie began her presentation, she said, "Now, even though the random number generator is active on your phone, you can see your location within the room."

The screen changed to a map of the United States, and she added, "Each of the locations are using the phone numbers of our people assigned to that specific location. There can be no case of a wrong number getting into your unique list of numbers.

"As you can see by the cities flashing yellow, these are the locations of the other operations that will commence at the same time, as I'm sure General Barker has told you," she finished. "Now, if there are no questions, I'll turn this briefing back to the General, and I'll wait for the next location to contact me."

Within mere seconds, the screen went blank, and Gene said, "Now, if there are no questions for me, I'll let each of you get with either Mike or Butch, and 1 will get to work making sure my end of the operation is on schedule."

Chapter 23

As the crowded conference room began emptying, Jim spotted Fabio and Tammie toward the rear. Waiting for the aisles to thin a little, he waved, signaling for them to go to the front of the room.

Spotting Mike and Jon toward the front, Jim headed for them. As he got there, they were discussing how to best use the limited number of NVGs.

"I hope you guys get it figured out soon, because if I'm not mistaken, they should be right outside in my car," Jim told them.

"Super," Mike replied. "I told Jon your suggestion regarding the motel, and he still believes he needs to put a set of NVGs at the back. Someone could crawl out of one of the windows at the rear."

"That's true," Jim agreed, nodding. "But there's the argument that since the hallways of the apartments are lighted, it would be a waste to put NVGs there. There are also arguments against using them in conjunction with the flashbangs.

"It's ultimately your decision, Mike," Jim said. "NVGs are definitely an asset under certain circumstances. Any external light lessens their effectiveness and could provide your target with an advantage due to his eyes being adapted to the ambient light. As I said, your decision, but really look at the specific lighting of each location.

"Consider street lighting, how headlights from traffic would affect the guy wearing the goggles. There isn't some all-encompassing rule," Jim finished. "I'd trust Jon since he's been to all the locations. But still your decision."

"Hey, guys," Tammie said as she and Fabio joined them. "I've got all the NVGs, and I've plugged them in with the USB cords to make sure they are charged before you take them to the field. I also picked up enough charging cords for the cars that will be taking the guys to their targets."

"Thanks, Tammie," Mike told her. "Did you turn them on and check them before you plugged them in?"

"No, I didn't think of that," she answered. "I figured since they are new in the box, they would need charging before they could be used."

"That's fine," Mike replied. "How long have they been charging?"

"Maybe thirty minutes," she answered.

"That should be enough for us to check them out," Mike said, nodding at Jon. "If they are working right now, we'll just take them and the cords with us and fully charge them in the cars we'll be using. Good for you, Jon?"

"Good as anything," Jon replied. "Most of those goggles use lithium AA batteries and come fully charged. But we've got several hours before we need them if they aren't."

"Okay, Jon," Mike told him. "You decide where you want the goggles and let me know if there's anything else you need. I'll give you a call when I pick up the flashbangs, and you can send your guys to meet me where we decided."

"Great," Jon said. "I'm going to give the goggles to one of the guys from each of the target locations and let them see how effective or valuable they are at each place. I'll let you know what I decide."

"Don't bother telling me," Mike told him. "You just handle it the way you and your guys think best. Tammie, could you take Jon to get the goggles, please?"

Waiting until they left, Mike asked, "What do you really think? I mean, Jon is a sharp guy, but he's not in charge of the guys who were assigned to Indy before he came up here."

"Do you trust him?" Jim asked.

"Yes," Mike answered. "It's just that he's making decisions regarding guys who aren't technically under him."

"Is he doing anything that would make you think he's favoring his guys over the other teams?" Jim asked pointedly.

"Not that I can think of," Mike admitted. "But it doesn't seem right to have him making decisions for the other teams."

"Is there any other team leader you would trust more?" Jim asked. "Not to bring it up, but Jon has more experience in these situations than you."

Pausing to let Mike think about it, Jim continued, "Jon is in charge of his team. Only his team. You are in charge of all the teams. But there's nothing to prevent you from letting him take care of some of the logistical issues, is there?"

"I guess you're right," Mike admitted. "Sort of like an assistant."

"Exactly," Jim said, nodding. "Not to mention, who's going to take charge if something happens to you? Who would be your choice to make sure the mission is accomplished?"

"Jon," Mike replied.

"I agree," Jim told him. "I haven't spent any time with the other team leaders, but I don't need to. I picked you to manage the teams for this operation against the TDA, same as I picked Butch for the Triad part.

"The only thing I'd advise is to make sure if something does happen to you, everyone knows Jon will be in charge," Jim added. "I realize I'm supposed to step in and take care of shit, but I've got to be concerned with handling the kids. Personally, I'm more concerned with them than the TDA.

"TDA is a known," Jim continued. "Pretty simple. Identify and immobilize. And they understand what you are going to do if they resist or fight back.

"I've never had to handle a bunch of children who are going to be scared to death, may or may not speak English, and don't have a clue about what's happening to them," Jim said, shaking his head. "Could be like herding cats.

"I'd like to have twice as many people as I have. Twice as many Mandarin speakers, preferably women, and bowls of ice cream to offer the kids, but I don't," Jim concluded. "Nor do I have someone who I can hand the problem to if something happens to me."

"Seems like that's why Gene doesn't have you in the field," Mike said. "Guys like me, Jon, Butch, we're expendable. Hell, I've always been expendable. That's what Marines are … expendable. Meat for the grinder."

"That's not the way Black Water works," Jim argued. "Black Water doesn't have a never-ending source of men.

You may have felt expendable in Vietnam, but you're not over there now. And I'm damn glad to have someone like Jon around to take over for you if necessary so I don't have to.

"Expendable?" Jim posed. "Not to this operation. And certainly not to me. Now, I'm sure you have more important things to do than listen to me whine about things I can't control. As my great-great-grandfather always said, 'Stop your whining you little pussy. Pull up your big boy pants and take care of the problem'."

Chapter 24

At 0530 the following morning, Jim walked into the conference room where the team supporting the raids was already assembled and watching the screen. The green icons denoting the field agents showed small clusters mere blocks from the targets, whose red icons showed they were where they were expected to be.

"Anything happening?" Jim asked as he walked over to one of the desks with a cup of coffee.

"Not much," the lady monitoring the icons' movements answered. "By the way, I'm Minzhu, named after my grandmother, but please just call me Minnie."

"Good morning, Minnie," Jim said, smiling. "I'm Jim."

"I know," she told him. "Nice to meet you. I'm here to mainly monitor Butch and his folks. I was probably assigned to be an interpreter if he, or any of his teams, need my services. I'm fluent in Mandarin."

"That was a good decision on someone's part," Jim told her. "Have any of the teams checked in?"

"All of them, sir," she answered. "The guys out by the farms are about a half a mile from their targets and reported that they are ready to proceed as soon as you signal."

"And the teams that are going to the apartments where the girls are?" Jim asked, watching five groups of green icons move and starting to see how the screen depicted the overall situation.

"They've checked in as well," Minnie answered.

"Any word from Mike and the flashbangs?" Jim asked.

"He met the airplane about two hours ago, then met with Jon and two other men," she answered. "Then he called back about thirty minutes later and said they had been delivered.

"Only one man at the apartment where the TDA targets are staying has come outside," she added. "He's currently sitting on the steps smoking a cigarette."

"Can you zoom in on that location?" Jim asked as he located the apartment on the screen.

"Sure," she told him as the screen showed a single red icon separate from the others.

"Do you know if the third TDA unknown from Colorado has been located?" Jim asked. "Can you zoom over to the motel where we think he is?"

"Sure," Minnie told him. "I'll get as tight of a shot as possible."

"Did you get any updates from Quantico as to the exact number of TDAs who came up from Colorado?" Jim asked as the motel site filled the screen.

"Yes, sir," Minnie said, picking up a sheet of paper. "There were fifty-two, and we've identified fifty-one."

"Can your computer give me a number of red icons on the screen at the motel?" Jim asked.

"Hang on a second," she said, tapping on her computer. "Now you know once the operation starts, it may be a little more difficult to zoom from one location to another."

"Can you do a split screen? Divide it into a section for each target area?" Jim asked as the red icons blinked off one at a time.

"I think so," Minnie said as a line of print appeared, as all the red icons reappeared. "Give me a couple of minutes. Could you please get me another cup of coffee while I'm setting this up?"

"No problem," Jim told her, picking up her empty cup. "Cream or sugar?"

"Just coffee," she answered as the screen began to divide. "I like my coffee same as I like my men."

"Black?" Jim asked.

"No … bitter," she said, smiling.

Jim laughed and headed for the table where coffee, tea, cream, sugar, and an assortment of sweet breads were laid out.

When Jim returned to where Minnie was finishing dividing the screen, he noticed that there were ten sections. Handing her cup to her, he asked, "Why are there ten sections? If I'm not mistaken, there are five farms and two apartments for the Triad operation, and an apartment and motel for the TDAs. What's the tenth screen for?"

"Just a second, and you'll see," she answered as each section filled with the areas Jim had mentioned.

Watching, Jim saw the tenth screen depicting the entire area. Then Minnie moved her cursor over each section and zoomed in and out.

"Great idea," Jim said, raising his cup to salute her. "Solves the switching from location-to-location issue. Thanks, Minnie."

"You're welcome," she replied, raising her cup. "Anything else?"

"Can you program it so I can see a running count of red icons in each area and the green ones?" Jim asked.

"You doubt by skills, grasshopper," she answered as she put her pointer on each section one after the other, and a line appeared at the bottom showing the numbers in either red or green. "Anything else?"

"How did you learn to manage all of this?" Jim asked as he mentally calculated the number of his men and targets.

"I worked with a lady back at Quantico when I was first hired," she told him. "Her name is Debbie, although I've heard her called Bracer by General Barker. Not sure what that means, though."

"That's a long story," Jim answered as he noticed several of the green icons in the sections showing the five farms begin to move toward the red icons. "I'll see if Gene, General Barker, wants to let you in on the secret. Or you can ask Debbie yourself."

"Not a chance," Minnie exclaimed. "That lady is brilliant on computers, but not necessarily social, if you know what I mean."

"That I do," Jim said, barely suppressing a chuckle. "That I do."

Jim left, shaking his head as he walked over to the desk where Amanda was sitting. "Have you made contact with all of the guys?"

"Almost an hour ago," she answered. "And with the ladies with the Triad apartments as well. Everything is up and running. That roaming frequency thing Debbie devised is unbelievably clear.

"I also made a couple of test calls to Butch and Mike," she continued. "Connection was the same as if their phones weren't part of the rotation of numbers."

"The lady is amazing … with computers," Jim acknowledged, taking out his phone and dialing.

"Mike," he said as it was answered. "Any problems?"

"Yeah," Mike told him. "I don't have a breakfast burrito. Can you get one delivered to me? With Pace Picante, please?"

"I'll call Uber or door dash," Jim answered laughing. Have you talked to the farmers?"

"I sent them to their pre-raid locations just a few minutes ago," Mike answered.

"I know," Jim told him. "I watched their little dots moving on my magical mystical screen. Guess that makes me all knowing and all seeing."

"Just be careful with all the action that's about to take place," Mike advised. "I'd hate for you to get excited and spill coffee on your pretty starched shirt, or worse yet, a nasty paper cut."

"I'm wearing a plastic bib and gloves, asshole," Jim retorted. "I'll be waiting for your call that all of the pieces are in place for the kick-off."

"Got it," Mike told him. "Expect it in about ten minutes."

"Sounds good," Jim replied as he saw Gene enter the room.

Chapter 25

"Good morning, sir," Jim said as Gene joined him, holding a cup of coffee and an apple fritter.

"Morning," Gene replied, looking at the screen. "Everybody ready?"

"Mike's going to call in a couple of minutes and I was just about to call Butch," Jim answered.

"Don't let me get in your way," Gene told him. "I just got word from Debbie that all of the other cities have reported standing by for the clock to tick down."

Jim stepped away and pulled out his phone. Dialing Butch's number as he walked back to where the coffee was, he picked up a bear claw and was just about to take a bite when Butch answered.

"Good morning, Jim," Butch said. "I've been expecting your call. The answer is yes. We're in position and ready to go. All the buses are just around the corner from the apartments, and we've accomplished communication and equipment checks. Anything new?"

"Nope," Jim answered not surprised that Butch had known the small details required before being ready to proceed. "Anything you need?"

"Luck," Butch answered. "Lots and lots of luck."

"Closest thing to luck I can offer is a box of Lucky Charms," Jim replied. "I'll be making a group call in just a few minutes to get it rolling. Let me know the minute you see anything that requires my attention."

"Got it," Butch affirmed him.

"Okay, time for me to get to work," Jim told him before hanging up and heading back to where Amanda was sitting.

Taking another look at the location of all the green icons, he asked, "Amanda, can you connect my phone to all of the other phones?"

"Sure," she answered. "When do you want it?"

"Now," Jim told her, taking out his phone and looking at the clock on the screen, which showed 0629:25 Eastern Standard Time.

As the clock reached 0630, Jim saw Amanda nodding and announced, "Go. Go. Go."

Almost immediately, they saw the green icons moving toward the red ones. "Looks like all we have to do now is wait," Jim said as the icons merged.

"Can you put all of the phones on speaker?" Jim asked while watching the screen.

"I can," Amanda answered. "But I don't think that's a good idea."

"Because?" Jim asked as the green icons began separating.

"Too many voices and background noise to really hear anything," she explained.

“You're probably right,” Jim agreed, nodding. “Okay, just put Butch on for now, please.”

A speaker located on the side of the screen came to life, and Jim heard him telling someone to get the older ladies first.

Seconds went by then Butch was heard saying, “Jackson, take care of the guy who’s coming toward you. Make sure you search him for any weapon. Especially looking for a knife. I don’t want him to cut his way loose after you get him in the back seat of the bus. Flex Cuff his hands and feet, then zip tie him to the seat in front of him.

“When you get the man and his wife, do the same. Try not to hurt the old lady. Just get her and her husband over to the bus. Put them together in the front seat on the left and zip tie them to the bar in front of them.”

“Sounds like they’ve got control of one of the farms,” Gene remarked, watching the screen.

“Minnie, can you zoom in on each of the farms?” Jim yelled across the room.

Without answering, the screen showed all five farms in greater detail.

As the green icons began moving separately toward where they had exited the bus, they heard Butch say, “Be careful with those kids. Put them two to a seat but leave an empty seat between each pair on the same side of the bus.

“Then alternate sides so there’s no one across the aisle from any other two,” Butch directed. “And Flex Cuff their hands … as loose as you can without allowing them to slip out.”

“Minnie, let me see the apartments where the girls are,” Jim asked, walking up beside her.

Both apartment locations on the screen zoomed in, and they watched as green icons spread throughout the buildings.

"Looks like things are going pretty well," Gene said, joining them with his eyes still glued to the screen.

"So far," Jim agreed. "But we're only in less than five minutes. I've been hoping we could get the buses loaded at each location within thirty minutes of starting. We've still got a long way to go."

"I just hope the other cities are going as smoothly," Gene said somberly. "At least I haven't heard from Quantico about any problems … yet."

"Hey, Jim," Fabio said, walking up to join them. "I'm sorry I wasn't here at the start. Anything I can do?"

"Yeah," Jim answered, still watching the screen. "Please get the car and bring it around front."

Turning to look at him, Jim asked, "Did you put our weapons in the car?"

"They are in the trunk," Jon answered. "I built a makeshift holder for the Bullpups so the stock will be sitting on the floor and the barrel held against the dash with some Velcro."

"That sounds good," Jim told him. "Make sure they are all loaded with one in the chamber. Same for the Glocks. Where do you plan on putting the spare ammo?"

"I removed the top of the center console and put all the G18 ammo in a cardboard box on the right side," Fabio told him. "I figured you'd probably need that more than the 12-gauge.

"The 12-gauge magazines are in the console also, in a box on the left side," Fabio explained. "Anything else?"

"Not that I can think of," Jim answered, turning back to the screen. "By the way, does the car have red and blues?"

"Yep," Jon told him. "This particular Suburban is on loan from the Secret Service. Bulletproof glass, run-flat tires, grill guard, everything except a toaster oven."

"Make sure you thank them when you return it," Jim said with a quick glance at him.

"I'll take care of getting everything ready to roll, and I'll be right back," Fabio told him as he turned to leave.

"I didn't know Fabio had friends in the Secret Service," Jim said, looking at Gene.

"I don't think he does," Gene replied, smiling. "But I do."

Chapter 26

Suddenly they heard Butch asking, "Are you telling me you've missed a guy? How the hell do you miss a full-grown man in a house where there are only three people?"

Hearing the response, Jim quickly dialed Butch's number and asked, "What the hell is happening?"

"The security guy at the furthest farm from me has disappeared," Butch told him. "His icon was in the house when we rolled in. I don't know, maybe he was up making coffee. Bottom line is he's not in the house."

"Got it. Hang on a second," Jim replied, putting his hand on Minnie's shoulder.

"Zoom in on that farm," he told her. "But stay far enough out to include the borders."

"I heard Butch lost a guy," Minnie said tapping on her computer keyboard. "I guess you think he's still on the property."

"Has to be," Jim answered. "He didn't go out the road where the buses came in, and he was in the house moments before the raid began. There's no other explanation."

"Here he is," Minnie said, pointing to a red icon a couple of hundred yards from the house.

"Send that to Butch's iPad," Jim told her. "Just make sure it's the exact same as on the screen right now."

"It should already be there," Minnie said. "But due to the size of the iPad, it's going to be a smaller image and harder to see with this much detail. He'll need to expand that section on the iPad to see where the guy is."

"Are you still on?" Jim asked Butch.

"I'm here," Butch answered.

"We've located your guy," Jim told him. "He's a couple of hundred yards north of the farmhouse. You'll have to expand the section of the screen on your iPad to get a good look."

"I see him," Butch said and then was heard to say, "Randy, he's north of the house. Couple of hundred yards. Look at your iPad. Upper right corner. Expand the screen."

"Good," Butch said. "Now, send someone out there. We can't have any of the adults get away. And make damn sure he doesn't have a weapon."

A couple of seconds later, Butch told Jim, "Looks like he's out in the damn corn field. I'll have to send a couple of more men to help him.

"It's like a jungle out there," Butch continued. "Damn rows of corn eighteen inches apart, seven feet tall, and as thick as hair on a dog's back. If he keeps running, it could be hours before we catch him."

"Use your iPad to direct your guy," Jim advised. "Better yet, have Randy direct him. Hell, he can see where the asshole is within three feet. Surely you can put someone right on top of him when he's within an arm's length. That's what all this damn technology crap is for. Use it."

"I understand," Butch countered. "If he's being chased on foot, he can hear my guy coming a mile away, crashing through all the stalks. It could take hours of running in circles."

"Hang on a second," Jim said, turning to Gene, asking, "Do you think you can get us a chopper pretty quick? As in right now?"

"I'll make a call," Gene said, stepping away.

"I see what Butch was trying to tell us," Jim told Minnie as they watched the red icon stop for a few seconds and then change directions as the green icon approached within a what appeared to be about fifty yards.

"Wish we had a team of good dogs that could run the man down," Jim said as the chase of the icons moved back and forth across the screen. "A couple of good German Shepherd Police dogs would make short work of that guy, even if he's a marathon runner. You can't outrun those dogs."

"Indiana State Police has one airborne about six miles south of the farm," Gene announced, walking up beside Jim and looking at the ongoing chase. "We can have him on his way as soon as you can give him directions."

"Did you get a frequency for us to give him directions?" Jim asked.

"Of course," Gene told him. "He's on VHF 122.8, or you can call him on the emergency frequency 121.5."

"Butch, does Randy have access to a VHF radio?" Jim asked.

"Not that I know of," Butch answered. "And neither do I."

"Well, shit," Jim said walking over to Amanda's desk. "Looks like I didn't think of everything you needed."

As Amanda glanced up, Jim asked, "Does your communication setup include VHF radio?"

"VHF, UHF, FM, or any others, including satellite communications," she answered. "What do you need?"

"I need to have VHF transmissions on frequency 122.8 sent to Butch's phone," Jim told her. "And if his guys at the farms can't hear the transmission, I need it relayed to one of his guys in the field."

"No problem," Amanda affirmed. "I can call Butch and have him tell me exactly which phone he wants to send the transmissions to."

"Can you send it to more than one phone?" Jim asked looking at all the green icons where Randy was trying to finish the raid on the farm.

"As many as you want," she answered.

"Good. Now, can you take the transmissions from those phones, convert them to the VHF frequency, and send them back to the chopper we're sending in?" Jim asked, hoping it was possible.

"Not a problem," Amanda answered. "As soon as I get Butch on the phone, I'll pull the phone numbers he wants to be on the VHF frequency from the random phone list that Debbie devised. That way, they stay tuned to the chopper's radio."

"Good," Jim said. "But let's not forget to put those numbers back in rotation as soon as we no longer need the VHF frequency on the phone."

As Jim returned to Minnie's desk and joined Gene, they heard the chopper pilot talking to Randy and asking for directions to the target.

"Sounds sort of familiar, doesn't it, sir?" Jim asked Gene as directions were provided redirecting the chopper every time the red icon changed directions.

"Do you mean where a guy on the ground is trying to direct an airstrike on a moving target to a flight of F4s, or the shaky voice of a helicopter pilot sitting on top of a vibrating machine?" Gene asked.

"You don't normally hear a Ground FAC, Forward Air Controller, have that unique sound on the radio," Jim answered. "Sometimes when the Army needs to move the FAC quickly, they'll provide a chopper to transport him. But he's normally sitting in the back without a radio."

"I remember," Gene said, nodding. "May have been years ago, but that shaking voice still wakes me in my sleep sometimes."

Chapter 27

Jim's phone rang, and he immediately heard Mike say, "Jim, we've got a problem at the motel."

"What's wrong?" Jim asked, walking over to where Amanda was monitoring the communications.

"There weren't as many TDAs in the rooms as we thought were supposed to be here," Mike answered.

"Put Mike on the speaker, please," Jim told Amanda.

"How many are missing?" Jim asked, looking at the screen view of the motel as he walked back to Minnie's desk.

"Just one," Mike answered.

"We sort of knew that," Jim replied. "That's not a surprise."

"I'm not talking about him," Mike told him. "There were a couple of them that weren't in their rooms."

"Wait a minute," Jim said, making a quick count of the red icons. "I see exactly the number we knew were there."

"Maybe so, but two, besides the unknown guy, were not in the room where they were supposed to be," Mike explained. "But we found them."

"Okay, what's the issue?" Jim asked.

"We had to take them down," Mike answered. "They must have left their phones in the room, and one of my guys saw them coming toward the back of the motel just before we were going to hit the doors.

"I had stationed one guy in the back with an NVG, and he was there to make sure there weren't any civilians around," Mike explained. "Anyway, he was ready to give the signal to hit when he noticed a couple of men coming toward the motel.

"Since he had the NVG on, he wasn't wearing his glasses, so we couldn't verify them as TDA," Mike continued. "Anyway, when they got fairly close, they noticed my guy and started acting strange.

"My guy told them to stop, but all of a sudden, they reached for their guns," Mike explained. "He had no option except to shoot them."

"That's why he was carrying a gun," Jim replied. "We knew going into this that there was a strong possibility that it would happen."

"That's not the worst part," Mike told him. "I'm guessing that the gunshots woke up most of the guys in the motel.

"As soon as the lights began coming on, we busted in the doors and tossed the flashbangs," he continued. "The shit really hit the fan. Those folks started shooting at the doors, the windows, every frigging where. Hell, my guys didn't have a chance to get into the rooms. We are damn lucky one of them didn't catch a stray bullet."

"What's your situation now?" Jim asked.

"I've pulled everybody back and I'm waiting until they either run out of bullets or realize there's no way out except through us," Mike answered.

"Shit, Jim," Mike finally said after a few seconds. "These are just kids. I don't mean like those Chinese children, but eighteen to maybe twenty-two.

"The guy who shot the two in the rear of the motel got a good look at the bodies and couldn't believe they were so young," Mike continued. "I'm afraid of what we'll find when we get a chance to hit the rooms again."

"I understand," Jim told him. "But these kids as you call them are the same ones who have been raping, murdering, robbing, or whatever other crimes they've been committing, across the entire country. Not to mention they're damn sure ready to take out all your people if given a chance.

"How many shields do you have?" Jim asked, thinking how he'd approach the situation.

"They've got enough for two for every room," Mike answered.

"Okay, here's my suggestion," Jim advised. "Give them a couple more minutes and move the guys with shields to either side of each of the doors. Don't let them expose themselves; just let the guys in the room see the shields. Maybe they'll understand what they're up against.

"Does anyone there speak Spanish?" Jim asked.

"I'll ask," Mike answered. "Hang on a second."

A minute later, Mike said, "We have one."

"Do you have a bullhorn?" Jim asked.

"Not a chance," Mike told him. "We didn't come here to talk to them through busted doors and broken windows."

"Okay, have the Spanish speaker see if he can get any of the TDA folks to talk to him," Jim said. "Maybe they all speak some English, and you can get them to leave their guns and come out."

"I'll give it a try," Mike told him. "Might take a few minutes."

"That's fine," Jim replied. "As we discussed, we don't want rivers of blood running down the streets."

"Speaking of streets," Mike said. "There have been several passing cars, and they definitely saw what was happening. I'm expecting the local law enforcement folks there at any minute."

"Make sure your bus with the ICE on the side is in front of the motel," Jim told him. "If you can block the drive coming in, do it. And have Jon there to talk to them. Above all, don't let them in.

"Tell them it's a crime scene under the control of the ICE, and active shooters are still involved," Jim said. "And get the two bodies in the back out of sight. Speaking of them, make sure your guy back there keeps an eye out for anyone else approaching the motel from the rear.

"We damn sure don't want some Lois Lane reporter trying to sneak in for her Pulitzer Prize winning expose," Jim continued. "Is there anything else we can do from here?"

"Not that I can think of," Mike answered. "Damn, I hate putting those guys in the middle of this FUBAR shit."

"I know," Jim told him. "Just try to keep them safe until the TDA folks decide to give up. We've got time, and we don't give medals for stupid."

Jim stepped back and looked down at his phone before saying, "Shit. Shit, shit, shit."

Gene walked over and put his hand on Jim's shoulder, saying, "We haven't lost anyone yet. Let's hope those kids come to their senses and realize death is their only option if they continue down the path they're on."

"Not to tell you how to do your job, sir," Jim finally said. "But I think it's time to let Quantico know what a shit storm this is becoming. Maybe they can give the Mayor, or someone else at that level, a call and ask them not to get involved.

"Not to mention, our covert operation has suddenly become seriously overt," Jim finished. "News people, TV cameras, pedestrians with iPhones wanting their faces on YouTube, grannies in their night gowns, Barney Fifes from across the city, and every Podunk village within the sound of the sirens blaring as they rush to the scene. This is going to look like Colorado times ten before we can get those guys, or their bodies, out of that motel. Especially if there's a trail of blood from each of the rooms.

"One more suggestion, sir," Jim said.

"What's that?" Gene asked.

"Maybe someone should let the real ICE know that they are about to be on the news," Jim answered. "At least we should give them the courtesy to have a little time to come up with some bullshit story to explain their presence at that motel.

"Hell, maybe we ought to dress the TDAs in white robes and sandals and tell the press they were members of an obscure sect of an ancient Mayan cult from Guatemala who were told by a blind three-legged dog that the world was ending today, and this motel was the only portal to the happy side of the afterlife and they came here to commit suicide," Jim said shaking his head. "What a shitstorm."

Jim looked at Gene and said, "I'm sorry, General. I didn't mean to sound so flippant. I apologize."

"No need, Jim," Gene said, putting his hand on Jim's shoulder. "After all these years of seeing the atrocities of man, I've learned that it's often gallows humor that seems to get us through the day. Take a quick break and then let's see if there isn't some solution to this shitstorm."

"I'm all right," Jim said as Gene removed his hand from his shoulder. "I've still got men out there who can't take a break. There'll be plenty of time when we wrap this up."

"Just keep your mind on the successes," Gene advised as Jim turned to walk away. "We haven't lost any of our people. And none have been hurt. We're still winning."

Chapter 28

Jim walked over to Amamda's desk and said, "Please put Butch back on the speaker."

"Butch," Jim said as his call was answered. "How are things going?"

"So far, so good," Butch said. "We got the runner and loaded everyone from that farm on the bus. They pulled out to go to the drop-off site a couple of minutes ago.

"Shit, Jim," Butch continued. "You should see what those kids look like. Randy said they were malnourished, almost skin and bones. Some looked like they had been beaten. I think you should let the folks where we're dropping them off know that they need some serious medical attention. These kids have been abused. Seriously abused."

"I'll take care of that," Jim assured him. "How's it going at the other farms?"

"At number two, I've numbered them according to their distance from my location, they are still searching for a couple of kids," Butch answered. "The wife was persuaded to tell us how many kids were living there, and there are two still missing."

"Have your guys looked out in the corn fields?" Jim asked. "Or maybe hiding in one of the hot houses where the marijuana is growing?"

"They are going through the hot houses as we speak," Butch answered. "But there are five of them and lots of places to hide. Speaking of hot houses, Randy asked if they could torch the ones from farm number one."

"Have them make one more sweep looking for anyone who may not have been in with the rest of the kids," Jim told him. "Our numbers estimating the number of kids could be off since none of them had phones, and I sure don't think there's any list anywhere. I'd hate to have a child hiding in there and burn to death.

"Not to mention if there were any new arrivals who may have been kept locked in the house until they were convinced their only option was to work with the other kids," Jim continued. "Thinking of that, make sure they search every room in the house. Look for a basement, an attic, even in the closets for hidden doors. Damn kids would probably starve to death before anyone would ever find them. And tell the other teams to do the same."

"Got it," Butch replied. "I'll make sure that's done before I let the teams leave. I'm still releasing the buses when it looks like we have all of those we knew about or found. I don't want to have all of them sitting in the bus zip tied and scared any longer than necessary."

"I agree," Jim said. "We can send a van to pick up any left behinds if we find any. How's it going with the parlor people?"

"Seems to be on schedule," Butch answered. "Maybe because we had a lady who spoke Mandarin. Anyway, all the kids are at least healthy, according to what I've heard so far."

"That doesn't surprise me," Jim replied. "Who'd want a skinny, bruised girl giving them a massage with a happy ending. Or even doing your nails. I'm sure those little girls get the best treatment possible. At least until they no longer suit their purposes."

"I'm beginning to wish I had never taken this assignment," Butch told him. "I'd have been perfectly happy never knowing what these people are doing to these poor girls … and boys. Unbelievable that a human would treat another person like this. Hell, even animals don't torture another animal for no reason."

"I know," Jim agreed. "But these folks aren't human. I've seen families reduced to selling their children in some of the countries around the world, especially the girls. As unfortunate as that family might be, I can't see doing that. But they live in a different culture, and what we find abhorrent is just a way of life in some of those places.

"I wish we could find a list of the clients and haul their despicable asses in. The people who abandon or sell their kids for survival are bad enough. But to take pleasure in sexually abusing a child, that's some sick shit," Jim finished. "Let me know when you find those missing boys and if you find any others."

"Hey, I just heard from farm three," Butch suddenly said. "They are taking fire from the farmhouse."

"I thought the houses were the first places your people were hitting, so we could get all of the adults," Jim said. "What went wrong?"

"I don't know yet," Butch answered. "Hang on a second, and I'll see if I can get any more information."

Jim walked over to Minnie and said, "Zoom in on the farm in the middle of the top row, please."

As the screen began to zoom in, Jim said, "No. I'm sorry. I need the one just to the left of that one."

Again, as the screen narrowed its view to just the house, Jim saw several green icons moving quickly toward where the cluster of red icons signified the location of the bus. Then he looked closely at the house to see if there was a red icon. Not seeing one, he looked at the bus parked merely yards away with the green icons on the opposite side of the bus from the house.

"Jim," Butch said excitedly, "The husband started yelling that the man in the house is his brother. He wants to be released to try and get him to stop shooting and come out."

Jim thought for a minute and answered, "Your call, but I think it's a good idea. Just don't let them remove his Flex Cuffs and have one of your guys holding him as they walk toward the house. And stay as far away as you think prudent. Can they see where the shots are coming from?"

"They say it looks like one of the front windows," Butch answered. "They are leaving the bus right now."

Jim watched the almost joined red and green icons moving slowly toward the house. When they stopped, he asked Butch, "Is the guy with the husband armed?"

"Oh, yeah," Butch answered. "Hang on a second. I'm listening to the husband talk. Damn, I wish I understood what the hell he was saying."

"He's saying to stop shooting," Minnie said, looking up at Jim. "He's asking his brother to stop shooting and walk to the door."

"He's telling his brother to stop shooting and come out," Jim relayed to Butch. "Hang on, I'm handing my phone to a lady here who speaks Mandarin. Her name is Minnie."

"Hello, Butch, this is Minnie," she said as she took the phone from Butch. "The husband, Li, also asked his brother to toss the gun out of the window."

"Is the brother saying anything?" Butch asked.

"I can't hear him if he is," Minnie answered. "Just a minute, the brother just told him that he would not be harmed if he'd just throw out the gun and step outside."

Another few seconds went by, and Minnie said, "I think he's going to throw out the gun. Li just said he'd wait for him where he was, but for his brother to keep his hands over his head."

"He just threw the gun out," Butch said. "Okay, the door is opening, and he's stepping out. What do you want me to do with him?"

"Tell him … hand me the phone, please, Minnie," Jim said, taking his phone from her. "Treat him the same as the rest of them. Search him and Flex Cuff his hands. I think he might behave better if you let him take the seat across the aisle from his brother, but that's up to you."

"Then what?" Butch asked.

"Turn him over to the Feds, just like the others," Jim answered. "And tell the Feds he attacked a uniformed officer."

"We aren't exactly uniformed officers," Butch told him as Jim watched Li's red icon join the others where the bus was located.

"He didn't know that. As far as he was concerned, you were. So, his intent was to attack a uniformed officer," Jim told him. "If he's an illegal, it's just another charge against him and another reason to deport him. Either way, he's been complicit in the operation of that farm and possibly abuse of the kids."

"I agree," Butch said. "Now, guess I'll send the guys back in to search the house. And I'll let all the other teams know about finding a shooter after we thought we had cleared the house. Damn, this could have been a disaster. Thank goodness you had a speaker there. That probably saved the brother's life. And could have saved my guys and maybe those kids on the bus. Damn!"

Chapter 29

"I wonder how much we don't know about these places," Jim said aloud, looking at his cold cup of coffee, walking over to where the coffee pots sat. "Shooters, missing kids, what else is going to come out of the woodwork to bite us in the ass?"

"You certainly didn't believe this would be a cakewalk, did you?" Gene asked as he poured Jim's cup with fresh coffee. "Hell, if we'd had perfect intelligence, we would have done things much differently.

"But we didn't," he continued. "Besides, there's no such thing as perfect intelligence. Everything in this world is in a constant state of change. Hell, even rocks change over time. Get worn down.

"Think about how much the world has changed in the last thirty years," Gene continued. "Hell, even more so the last ten. From rotary dial phones in your house to a hand-held computer, camera, and moving maps. All in a piece of plastic that can also phone home.

"In the last five years, the world has changed from personal contact to anonymous texting," Gene said, shaking his head. "That changes people. Makes them more disconnected from other humans. Sometimes I think it erodes the very soul of what it is to be human.

"After too much time seeing animated cartoonish characters replacing all of your friends' faces on your phone when you're talking to them, how can you understand the simple look of unhappiness or despair on a stranger's face?" Gene added.

"Don't get me wrong," Gene said. "That little piece of scientific wonder, the iPhone, has changed the world for the better. It has placed vast amounts of information right in the palm of your hand.

"You could be walking around the Somapura Mahavira in Bangladesh, be talking to your wife in Tucson, Arizona, sexting your girlfriend in Boston, and booking a flight from Los Angeles, California, to Austin, Texas, for a business meeting when you return to the US at the same time," Gene said as they moved aside for the hotel people who were wheeling in the breakfast buffet.

"What the hell is a Somapura Mahavira, or how did you know how to pronounce it?" Jim asked heading back to the front of the room. "And what the hell does any of that have to do with lack of intelligence in this operation?"

"Somapura Mahavira, one of the oldest Buddhist monasteries known. Been around since the eighth century," Gene answered. "Over one thousand three hundred years old. Slightly older than me.

"What I'm getting at, is that mankind has progressed so much scientifically. We have capabilities unimaginable just centuries ago, live longer, have better lives, fly to the moon. A million other things that would have been called a miracle, or witchcraft, a hundred years ago is commonplace today,"

Gene finished as they stood behind Minnie and looked at the screen. "But I think it's come at a cost."

"Everything has its price," Jim replied, putting his hand on Minnie's shoulder. "Anything new happening at the farms?"

"Not that I can tell. Been pretty quiet," Minnie answered. "I've done a little rearranging of the screen, as I'm sure you've noticed. I put the farms in the same order Butch has been using and put a number down in the lower left corner. That way I can immediately go to where you guys are talking about."

"That's a great idea, Minnie," Jim told her. "And I want you to know that I really appreciate you stepping in and helping with the situation at farm number two. You saved more than that brother's life. You saved the operation. And probably some of those kids from being injured or killed if a stray bullet had hit the bus. Thank you."

"You're welcome, Jim," Minnie said with a slight bow of her head. "I hesitated a little. I should have let Butch know that I was listening and made the translation sooner."

"You did great, Minnie," Jim repeated. "And please don't ever hesitate to let me know if you hear something that I obviously can't understand working with these Chinese people, but it could be important. Even if it's just a thought. Let me know.

"I just wish you could speak Spanish so you could interpret some of the stuff we're working with regarding the Tren de Aragua," Jim told her.

"I do speak a little Spanish," Minnie replied, looking up at Jim, smiling. "Burrito, tamale, queso, cerveza, tequila … should I go on?"

"Stick with the Mandarin," Jim said, shaking his head. Hell, I know more Spanish than you, and I don't know caca.

"Guess I better go check in on Mike," Jim said, turning away. "But seriously, you've done well today, and I'll make sure to let Debbie, or as you like to call her, Bracer, know. Maybe she'll invite you out for a wine and spill her guts."

Quickly turning back, he asked, "Say, Minnie. Do you think you could arrange the sections so that all the Triad operations are together and the TDAs are also?"

"If you'd really looked, it's that way now," Minnie answered. "But I've been meaning to ask you if you want me to remove any section once it's shut down?"

"I don't think so," Jim answered. "It's always possible we have to return after we drop off the bus loads. They may say something to the Feds that would require us to return. While we're talking about the sections, or the screen, is it being recorded?"

"Definitely, sir," she answered. "Both here at my station and at Quantico. One of the reasons it took me a little time to section the screen the way I did was getting permission from Debbie. It's technically hers, icons are generated by her, based on her programs, the photos of the land underlying the icons came from her, everything. So, I didn't want to make any changes without her permission."

"I understand," Jim said, nodding. "The lady can be a bit proprietary occasionally. And the gentlest dog in the pound can turn into a bitch if you try to steal her food."

Minnie smiled up at Jim and asked, "Are you calling …?"

"Oh, hell no," Jim said, cutting her off. "Do you think I would relish being a gelding? No way would I even contemplate implying anything of that sort."

Chapter 30

"Could you put Mike on the speaker, please?" Jim asked as he stood behind Amanda, looking at the motel section.

"You bet," Amanda said. "Everything good with Butch?"

"I believe so," Jim answered. "His operation is having some serious hiccups, but at least there hasn't been any shooting ... oh, wait. There's been a little shooting."

Amanda looked up at Jim with concern on her face and asked, "Was there shooting where Butch is?"

"Oh, no," Jim answered, seeing the concern in her eyes. "It was at one of the farms. But it pointed out some serious flaws in our making assumptions that proved invalid. I think Butch has learned a valuable lesson, as have I, about depending on information gathered weeks or months before an operation."

Pulling out his phone, he dialed Mike and when he answered asked, "Hey, Mike, what does up to your ass in alligators feel like?"

"I don't know," Mike answered. "You should ask Jon. That guy is up to his ass in alligators, reporters, FBI, CBS,

BS-NBC, WTF, and a partridge in a pear tree. I feel sorry for him. I bet he wishes he'd stayed skiing in Colorado. Do you think there's anything you can do to give him a little relief?"

"I'll see if Gene can make a few calls," Jim answered. "While we're waiting, did Jon get the bad guys to stop shooting at our folks?"

"For now, it seems," Mike answered. "It could either be to reload or they can't see any targets. Or they ran out of bullets. But for now, no bullets flying."

"How's the overall scene?" Jim asked. "Are there any unwanted curious people standing in the street with their iPhones?"

"Remember the number of folks in the pictures of the Colorado raid?" Mike asked. "We're rapidly approaching our own example of too many Indians *and* too many Chiefs. To make matters worse, Jon's got him a cop who seems to think he has a right to take over the scene."

"How's that going?" Jim asked, signaling for Gene to join him.

"More or less at an impasse," Mike said. "Appears as if Deputy Dudley Doright decided to block the entrance to prevent our block of the entrance, the bus, from leaving.

"According to Jon, and I can verify it because of Debbie's party line phone system, Deputy Dipshit threatened to arrest Jon if he didn't move the bus so he could exercise his authority to investigate the scene of a shooting.

"Then Jon threatened to arrest the Deputy if he didn't move his cruiser so he could transport his prisoners, who were under federal arrest, and not under city, county, or state jurisdiction, to a facility for processing," Mike told him, almost laughing. "Then, some panty-waisted reporter wannabe shoves his iPhone in Jon's face, saying he's recording everything Jon says, and demands to know his name and badge number.

"Jon looks directly at the guy and says, 'My name is Rexion. Let me spell that for you so you don't screw it up. R. E. X. I. O. N. Rexion. Sounds like wreck, car wreck, wreck-shun. Rexion, Hugh G. Rexion. My badge number is 3825968. Now, if you don't get your phone out of my face, you're going to need a proctologist to hit the play button for you," Mike said, laughing. "I like that guy. Didn't think much of him when we first met, but I like that guy."

"Yeah, maybe we should hire him in our PR department," Jim said. "What I really need to know is what progress have you made in arresting or subduing the TDAs.

"Give me some numbers," Jim continued. "On the bus, in the rooms, Flex Cuffed, dead, injured."

"As of right now, none are on the bus," Mike answered. "I thought it would be better to clear the area before we started loading it.

"That damn first shot at those guys who came up behind the motel sure started this FUBAR mess. First, it wakes everyone in the motel, then it alerts some driver passing, and he probably called 911, which alerted every first responder in two counties, as well as every news organization. Boy, if the General wanted a quiet get the money and run operation, I sure screwed the pooch on this one."

"Not your fault," Jim told him. "No one knew those two guys weren't in their rooms. No one knew a shooting war would break out seconds before the rooms were to be breached.

"Had those two guys been fifteen seconds later coming back to the motel, Jon's guys would have busted in and probably not a shot fired," Jim continued. "If they had carried their phones, we would have known where they were and improvised.

"Worst case under those circumstances, they would have heard the noise and rabbited," Jim said. "Maybe we would have never caught them. Maybe they would have led us to that unknown friend. Maybes don't count. Reality sometimes sneaks up and slaps us in the face just as we think we have control of the situation.

"Okay, back to my question," Jim repeated. "I need numbers. All I have are red and green dots. Some are moving, some aren't. Take your time and call me back when Jon has time to let you know."

Jim hung up and turned to Gene, saying, "I guess you heard all of that. I'm sure by now, even Quantico has seen what I'm sure is on every news channel in the nation. I'm not sure if trying to clear out the area wouldn't make matters worse."

"What do you have in mind?" Gene asked.

"I'll send Fabio over there, to the area behind the motel, and see if there is some way we can approach it without being seen from where the media is gathered," Jim answered.

"If there is, we can send another bus back there, load everyone, and leave," Jim continued. "I know we can't prevent the scene from being filmed, but we can at least leave the scene without having to fight our way through a sea of cameras, microphones, or possibly disapproving patrons. Hell, if we don't get this cleared up quickly, there may be some deportation protest group picketing the scene.

"I think Quantico should make sure we have a clear path out," Jim told him. "If we can get behind the motel, I sure don't want Jon to run into a roadblock anywhere between the motel and the drop off. Think that can be arranged?"

"I'll make a call," Gene told him. "Let Mike know we're working on an escape route and start planning on how to get the TDAs out of the motel as quietly and quickly as possible when the bus gets there, if Fabio can find a way."

"Got it," Jim said, looking around for Fabio.

Chapter 31

Finally spotting him eyeing the breakfast buffet, he called to him and motioned for him to join him. As he was coming over, Jim asked the people manning the desks, "Who's in charge of watching me?"

Pausing, he repeated, "Okay, folks. I know one of you is watching my little green dot to make sure I don't get surprised. Just as I know every one of you is assigned to monitor some other member of the teams. So, please, let me know who my guardian angel is?"

Just as Fabio joined him, he nodded to the lady who had raised her hand and said, "Thank you. What's your name, Miss?"

"Amber," the lady said bashfully.

"Good. Okay, Amber, do you know who is monitoring Fabio?" Jim asked.

"That's me, also," she replied. "The company figured since he's your driver, he'd pretty much be with you."

"Makes sense," Jim replied. "So, here's what I want you to do. Pay attention to Fabio for the next few minutes. I want you to be in constant contact with him and give him directions to where Jon is and find him a way to the back of that motel.

"Keep him at least four blocks from the motel until he gets to the street behind it," Jim told her. "Then let me know the minute he arrives. Think you can do that?"

"No problem, sir," Amber answered. "I've got a detailed street map of the entire Indianapolis area and can pretty much draw it out for him before he leaves here."

"Good," Jim said. "Let me know when you have a route planned. And I want you to monitor Fabio's phone. Just his. I don't want any random noise to prevent him from hearing you or not being able to contact you until he's at the rear of the motel."

Seeing her nod her understanding, Jim turned to Fabio and said, "I want you to check the approach and make sure we can get a bus and ambulance back there unseen from that crowd in front of the motel. You don't need to go all the way to the motel, just check the route."

"No problem, sir," Fabio replied, nodding. "Then what?"

"Use the same route and get your ass back here," Jim answered. "The way shit's happening, I want you here as soon as possible. Use your red and blues if you need to. But scurry back as fast as you safely can.

"I hope I don't need you, but this is fast getting out of control," Jim continued. "And, when you get back, come straight to me. Breakfast may be a little late for both of us this morning."

"On my way," Fabio said. "Amber, give me a quick call so we can make sure we're talking to each other."

"Yes, sir," Amber replied, reaching for her phone.

Jim walked over to where Tammie was sitting and said, "Tammie, I need you to find me another bus to transport the TDA folks from the motel. And I'll need an ambulance as well.

"Just get them headed toward the motel where Jon is and have them sit on the street behind it if possible. Amber will provide the location where they are to wait. Please let me know how long it's going to take to get them there."

"Got it," Tammie said. "I should have the bus in five minutes, and I've already got an ambulance on standby in case one of the locations needed it."

"Great," Jim said, pulling out his phone and dialing.

Walking back to Minnie's desk, he said, "Mike, I've got a plan. Let me know what you think and figure out how to make this work. Get Jon to start planning on a rear exit and let me know if you need anything else."

"What do we need to do?" Mike asked.

"Figure out how to get all of the folks out of the motel, including any bodies, as discreetly as possible," Jim answered. "My plan is to have a bus show up behind the motel where they are out of sight of the cameras, load everyone, including a couple of Jon's men to ride shotgun, and send them to the drop-off location."

"That's going to be some magic trick," Mike responded. "Jon says the crowd is doubling in size every five minutes. If you don't get something going pretty damn quick, we may completely lose control."

"Understand," Jim agreed. "Just talk to Jon and see what he thinks and if there's anything he needs."

"Give me a couple of minutes and we'll let you know," Mike said.

Jim walked over to Gene and said, "Sir, shit's going to *really deep shit* over at the motel. I need some help from the city or state to give us some help with crowd control. Is there anything you can do?"

"On it," Gene said, pulling out his phone. "How much time do we have?"

"I'm going to say ten minutes at the outside," Jim answered. "If we could get the city to set up some barriers at the entrance to the motel and provide some officers to man them, it would sure help."

Gene merely nodded as he turned to speak on the phone.

Jim walked back to Tammie and asked, "Any word on the bus?"

Holding up a finger, she said into the phone, "That would be great. Have the driver head toward West Washington Street and Hamblen West Drive. Same for the ambulance. Tell them to remain at least a half mile from the motel until we contact them."

"Yeah, Mike," Jim answered as his phone rang.

"Okay," Mike said. "Here's the plan. We'll take the bus and move it next to the motel. That will give us some privacy. Jon says since the motel is constructed of cinder blocks, he'll need a heavy saw with a concrete blade.

"He plans on cutting a hole at the back of one of the rooms so they can go out unseen by anyone on the street," Mike explained. "Since the guys are in adjoining rooms, we should be able to funnel them all through that room and out behind the motel. With the bus blocking the view from the street, it should work."

"Super. I'll have the saw coming," Jim told him. "Probably on the ambulance. Does Jon have control of the TDAs?"

"They're talking," Mike answered. "Jon thinks he can persuade the ones left standing to toss out their weapons and surrender. He may have told a little white lie or two."

"Tell him to say four 'Hail Mary's' and perform two acts of contrition when he gets home," Jim said. "I'll get back to you regarding crowd control as soon as General Barker gets off the phone. In the meantime, have Jon put a couple of his guys out by the entrance and hold their Bullpups, if they have them, so that the crowd can see them. If they don't have any, have Jon give one of them his Bullpup.

"I'm hoping the General can get you some help in the next few minutes so Jon can withdraw all of his people," Jim said as Tammie signaled him. "Gotta go. Let me know how it's going when you get a chance to talk to Jon. Biggest thing, get our people out of there with or without the TDA.

"If it comes down to the nut cutting, we can let that Deputy Dipshit take over," Jim said before hanging up.

"Yes?" Jim said, stepping beside Tammie.

"The bus will be there about the same time as the ambulance," she told him. "Will you need the driver?"

"No," Jim answered. Once he gets there, we'll arrange transportation back to wherever he needs to go."

Pausing, Jim said, "There's another thing we need."

"What's that?" Tammie asked.

"A concrete saw," Jim answered.

"Hang on a second," she told him, picking up her phone.

A couple of seconds later, she said, "The ambulance driver says he'll grab one from one of the fire trucks next door.

"Great," Jim said. "Please tell them to give the saw to Jon when they arrive. And be prepared for at least two dead bodies, and I don't know how many gunshot wounds. If they don't have room for the bodies, take the wounded. And follow the bus when it gets loaded.

"I really don't care if the wounded get first-class attention, just try to keep them alive, but don't even think about going to some hospital or emergency care place. They're just to follow the bus and give their patients to the Feds when they get there."

"I'll let them know," Tammie said. "But once in their care, I can't guarantee they won't do everything possible to save those people. Including going to the nearest emergency room."

"If that's their decision, just make sure they tell us as soon as possible. I may have to send someone from one of the teams that are finished to meet them there and provide security. Guess we'll have to adjust our plans if that happens."

Chapter 32

Almost ten minutes later, Fabio called saying, "I've found a good entrance for the ambulance and bus. It's about three hundred feet west of Hamblin West Drive, coming off West Washington Street. By the way, West Washington is the same as Highway 40.

"There are two drives just before they get to Hamblin; those two drives join about two hundred feet north of Washington," he continued. "Then a single road continues north behind the motel. The damn motel is only a couple of hundred feet from a runway at Indy International. I can't imagine trying to get any sleep with a damn airplane landing or taking off every five minutes."

"That's probably why the rates are cheap and even the TDA can afford a room," Jim replied. "Okay, once the bus gets to the road going north, what should the driver be looking for?"

"Okay, he's going to be able to see the motel parking lot on his right as he makes a small left curve," Fabio explained. "Then once he makes a hard right turn where the road he's on continues in a left circle, he'll be looking at the back of the motel.

"Another fifty feet or so, he'll pass a fairly large building on his right," Fabio continued. "If he turns right as soon as he can get past that building, he'll be at the back of the motel. He'll be where the motel makes a ninety-degree turn to the east. If Minnie is still watching me, have her drop a pin where I'm parked right now and send the bus here. They can see exactly what I'm talking about.

"If possible, that's where I'd park the bus until Jon calls for them," Fabio recommended. "It can't be seen from either the entrance to the motel off Hamblen or the one off Washington. Hell, with what little I saw of the crowd when I drove up the road I'm recommending, they seem so focused on the front of the motel that I doubt if they'd have seen me if I pulled into the parking lot. I don't think anyone east of the motel can see a damn thing back there. And they can take the same route back out and probably never be seen."

"Good," Jim responded. "I'll make sure Minnie marks the spot, and I'll let Mike tell Jon where to look for the bus. How long will it take you to get back here?"

"Shouldn't be much more than ten minutes," Fabio answered. "Traffic is still pretty light, except around a certain motel. It only took me ten minutes to get here, and I was driving pretty slowly, listening to the directions. So, less than ten minutes."

"When you get back here, leave the car in front and make sure the weapons are out of sight," Jim told him. "Thanks, Fabio. Maybe you'll have time for breakfast, if there's anything left except grapefruit and shredded wheat when you get here."

"That does it," Fabio replied. "I'm running the lights and siren. I want scrambled eggs, sausage links, an English muffin with blueberry jam, and a pretty lady to wipe my chin should I dribble my milk."

"I'll call Denny's if there's nothing left when you get here," Jim said, laughing. "Hell, if those assholes at the motel would hurry up and take care of business, I'll take you there myself."

"Minnie, make sure Tammie has all of the information Fabio just told me," Jim told her. "I don't know if Quantico is tracking them, but at least Amanda can get them on your phone so you can help them."

Jim turned, saying, "I'll go talk to Tammie and see if we can't get a phone number for the ambulance and the bus."

Standing quietly behind Tammie as she talked, he looked at the city map in the lower right corner, trying to get a bird's-eye view of the situation. When Tammie finally turned to him, he asked, "Do you have phone numbers or something we can use to track the bus and ambulance?"

"Sure," Tammie answered. "The ambulance was always part of the operation, so the driver and EMTs all have phones tied to our system."

Using the mouse on her computer, she pointed to a tight group of green icons, saying, "Right there they are, heading for the area Fabio just recommended.

"And, even if the bus isn't being tracked, he's following the ambulance," she finished.

"Understood," Jim replied, nodding. "Thanks for getting them so quick."

Walking back to where Amanda was talking to someone on her phone, Jim waited as she made a few adjustments to the section of the screen showing the motel.

As she finished her call, he asked, "Is there any way we can get the ambulance driver's phone number in the system?"

"I believe it already is, but I'll check," Amanda said, picking up her phone and watching the icon representing the ambulance heading for the area behind the motel. "Hey, Logan, do me a favor and give me a call back. Great. I'll be waiting."

No sooner had she put her phone down than it rang. Answering, "Hello?" she looked at the screen showing the ambulance approaching South Raceway Road. "Thanks, Logan. Hang on a second. There's a guy here who would like to talk to you."

Handing the phone to Jim, Amanda quietly asked if there was anything else she could do. Holding his hand over the phone, he asked, "Can you put Logan on a private line with Mike and Jon?"

"Sure," Amanda answered. "I'm a pretty good switchboard operator. Plug lines A, B, and C together. No problem."

"Please do and let me know when I can talk to all three at once," Jim said, taking his hand off the phone.

"Logan," he said as Amanda nodded and started tapping on her computer. "This is Jim Lashley. What I'm trying to do is get you connected with the team leader of the TDA problem as well as the guy directly involved with the motel cluster fuck where you are heading."

"What can I do to help?" Logan asked while watching for road signs and listening to Minnie tell him where to go.

"I don't know how much Tammie told you, but you're about to enter a circus of cops, reporters, and who knows how many folks with nothing to do but watch the cops and reporters," Jim answered. "I'm hoping I can keep you out of the middle of this three-ring circus. A major part will be coordinating between you and the two people I mentioned."

Just as he was about to go into greater detail regarding any wounded, Amanda held up three fingers signifying the three phones were in contact with each other.

"Hang on, Logan, let's see if we've made contact," Jim told him. "Mike, do you hear me?"

Hearing he did, Jim asked, "Jon, do you hear Mike and me?"

"Great," Jim said. "Now, Logan, do you hear both of them?"

"Sure do," Logan said.

"And I hear him, too," Mike replied.

"Me, too," Jon joined in.

"Okay, guys, here's what I want the three of you to do," Jim began. "I want Jon to have someone behind the motel when you get there, Logan. Once Minnie tells you that you're close to the building where you'll cut across to the rear of the motel, start a running conversation with him so there's no chance of not parking where he wants to cut into the building.

"Mike, I want you zeroed in on this," Jim said. "We can get back to the apartment issue when we get this resolved. Just watch the moving dots and be ready to jump in if you see a problem coming.

"Jon," Jim said. "Once you get your bus relocated to block the view of the motel rooms you'll be using, make sure your guy at the back knows exactly where to make the cut. Knock out a window or something. We definitely don't want to cut into a room where the Mayor's wife is entertaining her Sancho."

"If it suits you," Logan said. "I normally handle the saw when there's been an accident. If it would free up your men, I'll make the cut. Just show me where it is. Please look for any electrical boxes or conduits before selecting which room you want to enter."

"Fine with me," Mike said. "I'm glad to turn this over to Jon. If either of you needs something, I'll be listening."

"Good with me," Jon chimed in. "I'll send a guy back to clear the area right now. It may mean we have to select a different room, but that would be preferable to finding out it won't work when you get here. I'll just knock out one of the rear windows and let him start there, looking for the best location if that area doesn't look promising."

"Next," Jim said. "Jon, please have a couple of guys back there to get the bus parked and escort the driver somewhere out of sight. One of your guys will be driving the bus to the drop-off location.

"If everything goes as planned, this shouldn't take more than five minutes from when Logan turns off of Washington Street," Jim explained. "Gene is supposed to be getting some barriers for the street, and a couple of officers to man them. But until they get there, make sure nobody approaches the motel.

"Especially once Jon relocates his bus," Jim continued. "I'd advise waiting until you hear Logan getting close to the turn off on Washington. This is where all of you need to be talking. I could watch the screens here, and Mike could watch the green dots move, but there's nothing that can replace a running dialogue."

"How far are you out, Logan?" Jon asked.

"Maybe five, maybe six minutes," he answered.

"I'm sending my men to the rear," Jon said. "We'll direct you to the spot."

"Have you started moving the TDAs to the room where you'll make your exit?" Mike asked Jon.

"Yeah, but I'm going to check the rear like Logan suggested to make sure we can get out from that room," Jon answered.

Jim listened for a few seconds and then handed the phone back to Amanda. "Guess we'll just have to wait and let them do their jobs."

Just as Jim was turning to see if he could get their conversations on the speaker, Gene walked up and said, "The Governor approved two state police cars, they should be on scene about now. The city decided they didn't have any barriers they could loan us. Nor could they spare any of their officers."

Jim spun back to Amanda and said, "Get me on the line with Logan and the other two."

As soon as she had made the connection, she handed Jim the phone, and he said, "Listen up, guys. The Indiana State Police should be arriving any time now if they aren't already out front. Jon, send someone to take a look. If they're there, I'd recommend getting your bus in position to block any views from streets."

Two seconds later, Jon said, "They just pulled up and are trying to get through the crowd. I'm sending a guy out to move the bus since it appears no one is paying attention to us right now."

"Thanks, Amanda," Jim said, handing her phone back. "*Now*, I guess we wait."

Chapter 33

Just as Jim started over to where the breakfast buffet was set up, his phone rang. "Hello," he said as he picked up a slice of crisp bacon.

"Jim, this is Butch," Butch said excitedly. "I've got a problem. Major problem."

"What's the problem?" Jim asked, shaking his head.

"I've lost one of the supervisors, or madams," Butch answered.

"What do you mean you've lost her?" Jim said, tossing the remnants of the bacon in the trash can and heading back to look at the screen.

"One of my people had just cleared out one of the rooms, and the Chinese lady who was there with the girls suddenly grabbed her and put a knife to her throat," Butch explained.

"Which apartment?" Jim asked, looking at the two apartment sections.

"The one up on Meridian Street," Butch answered. "Just north of twentieth street. On the east side of Meridian."

Tapping Minnie on the shoulder, Jim nodded to the screen and quietly said, "Please expand section six, lower left side."

"Got it," Jim told Butch. "What the hell happened? Why wasn't that woman searched?"

"Jewell, the lady who was bringing the little girls out, was just beginning to search her," Butch explained. "The Chinese lady pulled a knife from under her dress, grabbed Jewell around the neck, and started screaming.

"She pulled Jewell and four of the little girls out of the door and told everybody to stay away," Butch exclaimed. "Then she forced Jewell to head to a car and made the girls follow them.

"One of the guys started to take a shot at her, but he was afraid he'd hit either Jewell or one of the kids," Butch continued. "The lady forced Jewell into the car and made the girls get in the back.

"Then, they started the car and headed out of the apartment parking lot," Butch finished. "Dale, the team leader there, said there was no way they could chase them in either the van or the bus. What do you want me to do?"

"Did the Chinese lady still have her phone?" Jim asked.

"I think so," Butch answered. "Dale didn't see it here."

"What kind of car was it?" Jim asked.

"Some sort of SUV, Range Rover or something like that," Butch answered. "Brown."

"Okay, go ahead and get everyone else loaded on the bus," Jim told him. "I'll call you when I figure out what we'll do."

Jim hung up and asked Minnie, "Can you rewind the apartment tape?"

"Yes, sir," Minnie said. "How far back do you want to go?"

"Just rewind slowly until you see a red icon leaving the apartment," Jim said.

Watching the scene reverse, Jim suddenly said, "There! Now go forward slowly. I need to see which way that dot goes. That's got to be our rabbit."

He watched the car come out of the parking lot and turn right on Meridian Street heading north.

"Can you follow that car?" Jim asked as he saw Fabio coming in.

"Of course," she said. "But I can't tell you where she is right now. I'll have to fast forward and hope she's still on the screen."

"Zoom out and then forward until you get to real time," Jim said, motioning to Fabio.

Watching the screen compress until the red icon came back into view a mile or so north of the apartment, Jim said, "Keep that dot on the screen and focus on it. Forget about the apartment. Just follow that dot, and me, when Fabio and I leave. I want you to direct me to wherever that car is going."

"Yes, sir," Minnie said. "Do you want me to call your phone so I can tell you?"

"Yes, ma'am," Jim told her as he told Fabio to follow him.

"We've got a runner we need to catch," Jim said as they headed out of the conference room. "Now's the time for the red and blues."

As they got to their car, Jim's phone rang. Answering, Minnie said, "She's still going north on Meridian. I suggest going north on Pennsylvania. I'll have an intercept for you when you get a few blocks north."

"Okay, don't hang up," Jim said as Fabio turned on the lights and sped out of the hotel onto Pennsylvania. "Just keep track of her and me. I'm putting you on speaker so Fabio will hear your directions."

"Okay, you're going to make a left on East North Street. It should be about half a mile in front of you," she told them. Then East North Street will take you to Meridian. It should be the first intersection after you turn.

"Then you'll be on Meridian," Minnie said. "She's still going north on Meridian."

"Thanks, Jim said as Fabio blew past several cars with his siren blaring.

As Fabio drove, Jim picked up one of the Bullpups, checked the magazine, and made sure a shell was chambered.

Next, he took one of the Glocks and checked the magazine and the chamber for a round. Putting the Glock into the back of his jeans, he fastened the Bullpup against the dash with the Velcro strap Fabio had installed.

As Fabio skidded, turning left on North Street, he checked the other guns and then asked, "Minnie, how far is she in front of us?"

"Maybe six or seven miles," she answered. "It's hard to judge because I don't have a scale on the screen. So, I could be a couple of miles off. Probably it's further than I'm thinking."

"No problem," Jim said as Fabio whipped around the corner, joining Meridian. "You're doing just fine."

Chapter 34

Just as they were crossing under I-65, Jim's phone signaled an incoming call. Glancing at the caller ID, he saw that it was Mike.

"Hand me your phone, Fabio," Jim said. "I need to call this guy back. My phone's still on speaker, so you can hear Minnie."

Dialing, he said, "Make it quick, Mike. I'm sort of busy right now."

"We got the folks out of the motel, and they're headed for the drop-off," Mike told him. "The ambulance took three wounded TDAs, and we put the two bodies in the bus. Sort of a reminder to them what would happen if they screwed with us."

"Okay, stay on top of it," Jim told him. "I assume you put a couple of men in the ambulance."

"Of course," Mike answered. "We also flex-cuffed their hands and feet. Then we strapped them down nice and snug. So far, Logan is content to continue to the drop-off. But he told Jon that if it appeared either of the TDAs were having a serious issue, he'd have to go to the nearest emergency room."

"Okay," Jim replied. "Make note of this number and use it if you need to talk to me again. But try to take care of everything until I get a chance to call you again."

"Got it," Mike said. "What's happening?"

"Not now," Jim replied. "I'll fill you in when I get this resolved."

"The lady just crossed Kessler Boulevard West," Minnie told them. "You should be about four miles behind her. She's still headed north."

"Hey, Jim," Amanda said on Minnie's phone. "That car you're chasing just made a phone call."

"Did you hear it?" Jim asked.

"Yeah, but it was in Chinese," Amanda said. "So, I couldn't understand a word."

"Did you record it?" Jim asked as Fabio rode the bumper of a car that was blocking the lane.

"Yeah, what do you want me to do with it?" she asked.

"Take it to Minnie and let her translate it," Jim replied. "She can let me know if it's important. By the way, did you get the number she was calling?"

"Yes, sir," Amanda answered.

"Call Quantico and see who she was calling," Jim told her. "And, if possible, get an address connected to that phone."

"I'm on it," Amanda replied.

"She's about to go under 465," Minnie told them. "You're a little over a mile behind her, and she's still headed north."

"I see the signs for 465," Fabio said as he passed the car that had finally decided to pull over and let him by. "At least now I can speed up a little. Let's see what this piece of shit can do."

Jim watched as the speedometer passed one hundred and rechecked his seatbelt. "We should be able to spot her in a minute … if we're still alive."

"No worries, mate," Fabio said, glancing at Jim. "I'm a professional."

Just as they passed beneath 465, Jim said, "Up there, looks like a brown SUV. Get on her ass."

Weaving through the cars between them, Fabio sped to within yards and slowed to match the brown SUV, saying, "That's got to be them. I see two women in the front seat and… looks like kids in the back."

"I agree," Jim said. "But it doesn't look like she intends to pull over."

"I'll give her a bump and see if that works," Fabio said, inching closer to their bumper. "Just a tap."

"Just remember, we've got one of ours in that car," Jim said as they made contact.

Suddenly, an arm extended out of the front passenger window.

"Look out!" Jim said. "Gun!"

Before Fabio could back off, a bullet hit the top of their windshield, making a spider web pattern in the broken glass.

"Shit," Jim said. "Where the hell did she get a gun?"

"Hell if I know," Fabio swerved to the left. "What do you want me to do now?"

"Do you know what a pit maneuver is?" Jim asked.

"One of my best moves," Fabio said, staying to the left of the SUV.

Jim yelled, "Minnie, get an ambulance headed our way. You know where to send them."

"Do it now," Jim told Fabio. "Try not to roll them over. Just make them skid."

"I'll do my best," Fabio said. "No promises. Some of those SUVs are a little top-heavy."

"I know," Jim said. "And then try to block the driver's side doors with your grill guard when they're stopped. I don't want to have to watch both doors when I get out. Once you stop, grab the shotgun, get out, and stay behind this car by the front fender."

Fabio accelerated until his front fender was just past the rear of the SUV and then swung right into the car.

As they hit, the front of the SUV started coming left across the front of their car. Just as they thought it was about to stop, it suddenly rolled right and tumbled over coming to a stop upside down directly in front of them, blocking the road.

As Fabio rolled up against the car, Jim unbuckled his seatbelt and pushed open his door, saying, "Don't shoot unless you see the woman shooting at me. And make sure you don't hit me or their car."

Chapter 35

Rushing up to the overturned car, Jim pulled out his pistol and ran around to the driver's side. Looking in, he saw Jewell hanging upside down with blood pouring from her head.

Across the seat, he saw the Chinese lady lying on the headliner, her head almost touching Jewell's.

Finally, he managed to open the door. He tried to hold Jewell up and unbuckle her seatbelt, saying, "Fabio, come give me a hand. I've got to get her out so we can see if the Chinese lady is still alive."

As Fabio reached the door and held her up as Jim finally succeeded in freeing her seatbelt, he said, "I think I smell gas, Jim."

"Yeah, I smell it, too. Let's get Jewell away from the car," Jim replied as they extracted her unconscious body and placed it just off the road. "You go move ours back in case this one catches on fire.

"I'll get the kids out of there, and we can open the rear of the Suburban to hold them until an ambulance gets here," Jim continued as he struggled to get the rear door open.

"And ask Minnie where the ambulance is and how long until it gets here."

As Fabio backed the Suburban away from the overturned car, he moved it to the side of the road with the rear facing Jim to make it easier to transfer the injured kids to the rear compartment.

Killing the engine, he grabbed Jim's phone from where it lay on the passenger seat and said, "Hey, Minnie. Where's the ambulance? We've got several injured kids here, an unconscious or dead Chinese lady, and one of our people with at least a head injury."

"There's a hospital about half a mile north of you," she answered. "They should be there in the next couple of minutes."

"Thanks," Fabio said as he joined Jim, pulling the kids from the car. "Ambulance is coming. Couple of minutes."

"Can't be too soon," Jim said as he picked up Jewell and headed for the Suburban.

"This little accident is going to start getting some unwanted attention," he said, laying Jewell in the rear seat.

"I know what you mean," Fabio said, pointing his head to the cars coming up behind them.

"Let's get these kids loaded and be ready to get the hell out of here," Jim replied as he carried a little girl with blood running down her face to the open rear door and laid her gently inside.

"Hey, what the hell is going on here?" a man asked, walking up to the side of the Suburban.

"Just stay back," Jim told him, turning to look him in the eye. "This is an active accident scene involving a criminal organization."

Taking one step toward the man, Jim added, "And there's a good possibility their car could catch on fire. So, please, return to your car. There's an ambulance on the way, and they need to have access to get these folks to the hospital."

"Wait a minute," the guy replied, not moving. "You're ICE. Is this something to do with that deportation crap I've been seeing on the news?"

"Get the hell back to your car, Mister," Jim said over his shoulder as he turned back to where Fabio was bringing another kid to the Suburban. "Now! You're interfering with an officer in a federal operation. Get the hell out of here!"

Fabio placed the little girl he was carrying into the car, stepped to the open passenger door where he had left his Bullpup, spun around facing the man, and said, "Sir, you've been asked to step back. You've been told you are interfering with a lawful operation. Now, for the last time, get your skinny ass back to whatever car you came in. And do it now before I cuff you and toss your butt in my car."

As Jim was carrying the third girl to the Suburban, a siren was heard coming from the north. Laying the little girl in the back of the Suburban, he looked at the guy still standing and watching, and said, "Sir, please. The ambulance is almost here, and there are still a couple of people we need to get out of that car. Please. Please. Go back to your car so we can get these people to the hospital."

"Why was your car up against theirs?" the guy asked, still not moving. "Did you wreck their car?"

"Okay," Jim said, walking over to the man and putting his face an inch from his, saying, "Last time, scooter. Get the fuck out of here. If you're not back in your car when the ambulance gets here, I'm going to snap your little pencil neck like a chicken bone."

Spinning away and passing Fabio as he was putting the last girl in the Suburban, he said, "Let's get the old lady out.

I'm not sure what her condition is, but I want all the people out of that car so the ambulance can take them off our hands as quickly as possible.

"And I really want to get the hell out of here before some damn news crew shows up," he continued as he knelt, reached into the car, and tried to pull the limp body of the Chinese lady out.

"Let me give you a hand," Fabio said, putting his hands under her shoulders. "Minnie's been calling you, but I didn't think you'd want to talk to her right now."

"I'll call her back when we get this lady to the Suburban," Jim said, nodding. "Let's put her in the front passenger seat."

As they were setting her in the seat, Fabio said, "That guy doesn't seem to want to leave. What do you want me to do?"

"Hand me the phone," Jim said as they got the lady's limp body settled in the seat. "Go ask him one more time. Nicely."

Holding the phone to his mouth, he saw Fabio step up to the man, get in his face, say something, and then roughly shove the man backward with both hands. As the man hit the ground hard on his butt, Fabio turned, looked at Jim, shook his head, and slightly raised his hands signifying 'what else could I do?'.

"Hey, Minnie," Jim said, turning away and walking to the side of the road. "What did you need?"

"We found the address for the phone the lady was calling," she answered. "And I listened to their conversation."

"What did the lady say?" Jim asked as the ambulance slowed beside the overturned car.

"She was screaming that their apartment had been raided and that she was coming to his house," Minnie told him. "Then the man on the phone told her not to come there. Then she said she had nowhere else to go. Then he said he didn't give a shit, just to go somewhere else and then he hung up."

"Have you looked up the guy's address?" Jim asked. "I'm betting it's somewhere up this road."

"You're right," Minnie replied. "His address is about ten miles north on Meridian and east a mile or so in a residential area. What do you plan now?"

"I'm thinking about that," Jim answered. "Probably go pay the man a visit and see what his involvement is. Did you happen to get a name to go with the address?"

"It's listed as owned by Ming Li," she answered. "Let me know if I can do anything else. Oh, by the way, Mike has been calling asking what you want him to do."

"I'll give him a call when we get these people loaded in the ambulance," Jim said, watching the EMTs putting Jewell into the rear. "For now, find me a route to the guy's house. I'll give you a call when I'm sure of what I'll do next."

"Can you get everyone in this ambulance?" Jim asked, walking up and looking inside at Jewell's still form.

"Not a chance," an EMT told him. "We're going to get the older lady loaded and head for the hospital. We've taken a quick look at the kids, and they are banged up a little, but nothing life-threatening.

"There's another one on the way, and they'll bring them to the hospital," the EMT said as they loaded the Chinese lady into the ambulance.

"How's that one?" Jim asked, pointing to Jewell.

"She's had a pretty bad hit to her head," he answered. "Nasty cut, that's where all the blood is coming from. But I think she'll survive if I can get her to the hospital before there's any swelling in her brain."

"And this one?" Jim asked, nodding toward the Chinese lady.

"I don't think so," the EMT replied, shutting the rear door. "I couldn't find a pulse, but I can't make that call. It's up to the doctor when we roll her in. Now, sir, if you'll step back, we'll do our best with both ladies.

"The other ambulance should be able to take all of those little girls," he continued, heading for the driver's door. "They can also let you know where they are taking them."

"You're not going to the same hospital?" Jim asked as the motor started.

"Not positive," he answered, shutting the door. "We're heading just up the road to Health North. They will probably meet us there, but I'll let them tell you for sure."

"What now, sir?" Fabio asked as they watched the ambulance head north.

"I guess we'll have to wait for the other ambulance to get the kids, and then we'll take a drive up north," Jim said looking to where the obnoxious man had formerly been sitting on his ass. "Looks like you finally convinced our friend to behave. But I'll bet he won't be quiet about it."

"At least he didn't have a phone stuck in our faces," Fabio said, shaking his head. "I wouldn't want that on the nightly news."

"Oh, grasshopper, you are so naive," Jim said as they heard the other ambulance approaching. "Maybe he didn't. But how many other cameras do you think were there filming your little altercation? I'll give you a hint … how many hairs are on a dog's butt?

"Don't know?" Jim continued as the ambulance arrived. "Neither do I. But I'm sure there are more than a handful.

"No, my friend, you're about to be a national sensation," Jim told him, watching the ambulance back up to their

Suburban. "Maybe the *All Elite Wrestling*, the AEW, will name a takedown move after you. Maybe call it the *Full-frontal Fabio*?

"You better get yourself a good agent," Jim continued, waiting for the EMTs to get out. "Don't let them make millions off a wrestling move you've spent years perfecting."

"Excuse me," Jim said as they began to load the girls. "Where are you taking them?"

Hearing they were headed to the same hospital as Jewell; he thanked them and walked back to the overturned car with Fabio by his side. "I'm going to take a quick look in the car," Jim said, bending over. "I'd like to find the lady's gun. Maybe they can trace it."

A minute later, he stood up holding the gun and a phone, saying, "I bet this is her phone. Maybe we can use it to make an appointment."

Watching the last kid get loaded and the ambulance pull away, Jim called Minnie and asked, "Are you still watching that phone the lady called?"

Quietly telling Fabio they were headed out, he got into the Suburban and asked, "Has there been any activity on it? Is it still at the address you have?"

Hearing there hadn't been any activity, and it was still there, Jim said, "That's where we're headed. Please give us directions."

Pulling his seatbelt on, he said, "Well, Fabio. Let's go see why the gentleman didn't want to see the nice lady. Head north until Minnie gives us a turn."

Chapter 36

Heading north on Meridian, they were less than a mile behind the speeding ambulance. A couple of miles later, they watched as it pulled into a hospital on the left side of the road.

"We should stop by on our way back to the hotel," Jim said as they passed, "See how Jewell is doing."

Jim looked at the phone he had taken from the wrecked car and opened it, saying, "Minnie, give me the number of the phone the lady called."

After giving Jim the number, she asked, "What's your plan? Are you going to call him?"

"That's the idea," Jim said, dialing the number. "Just listen to whatever he says and then tell me after I hang up."

"What are you going to say?" she asked.

"Nothing," Jim answered. "I just want him to think the lady is still on her way."

As the phone began ringing, Jim could hear an agitated voice yelling in Mandarin. Finally, after only a few words, the phone went dead.

"What did he say?" Jim asked, hanging up the phone.

"Something like 'I told you not to come here, you stupid cow. Don't ever call me again. I'll find you when I'm ready'," Minnie told him.

"Good enough," Jim said. "At least he thinks she's still alive."

"What good does that do us?" Fabio asked.

"Maybe nothing," Jim answered as the road curved to the right. "How far are we from the address, Minnie?"

"About four or five miles before you go east on 151st street," she answered. "Then another mile or so to Oak Road, on your left. Then, in about a quarter of a mile, you'll come to a subdivision on the right. The entry is on Maple Ridge Drive. I'll keep giving you directions until you tell me you see the address.

"Gene is standing here with me and wants to know what you plan to do when you get there," Minnie told him.

"Knock on the door and let Debbie's facial recognition glasses do their job," Jim answered. "Let's see if this guy is in her databank of illegals. Maybe he's one of the Triad leaders here in America.

"I know I'm supposed to be standing back there making sure this operation is going smoothly, but I'm afraid if I don't take this opportunity right now, this guy could be long gone and we'll never know his role in the sex trafficking thing," Jim continued as they crossed over a major intersection with 136th street. "I should be back within thirty minutes, and we'll at least know who he is."

"You need to get in the far-right lane," Minnie told them. "Start watching for the exit to 146th Street. You need to take it and stay on the service road until you reach 151st."

"Got it," Jim answered after seeing Fabio start merging with the traffic to his right.

"What's your plan when we get there?" Fabio asked, finally managing to get into the lane he wanted.

"Knock on the door, introduce myself, and ask if he's heard the good word," Jim answered, pointing to the large exit sign for 146th Street.

"You forget, you're wearing a uniform with a huge ICE across the front and back, a bulletproof vest, and a black tactical helmet with a face shield," Fabio replied. "Do you really think this guy is going to believe you're some Jehovah's Witness here to save his child-abusing soul?"

"They wear black suits; this is a black suit. Okay, maybe it's just a black jacket. Okay, it's a black tactical vest. Close enough for the girls I go out with," Jim told him as they made the right turn onto 151st Street. "I just wish I had some pamphlets to leave with him."

"I'm serious, sir," Fabio said as Minnie told them to turn left in a little over a mile on Oak Road. "What do you really plan to do if he opens the door?"

"First, I'm hoping he thinks it's the lady who was calling him, saying she was coming," Jim answered as they went through a roundabout. "When he opens the door, I plan on shoving my Glock under his chin and stepping inside with him."

"What do you want me to do? Sit in the car?" Fabio asked as they turned left on Oak Road.

"About a thousand feet, on the right, Maple Ridge Drive," Minnie reminded them. "Then, about a half mile, you come to Oak Bluffs on your right. The address is at the end of the cul-de-sac, which makes a circle. House is at the very end."

"Thanks, Minnie," Jim said as he pulled his Glock from the back of his jeans and rechecked it.

"I want you to be six inches behind me with the barrel of the Bullpup visible over my left shoulder," Jim answered as they turned on Maple Ridge. "Then, I want you to follow me in and clear the left side of the room.

"I really don't expect the man to be armed, but I wouldn't bet against a bodyguard or two backing him up," Jim continued. "If you see anyone making a move for a weapon, take a step to the left and take them down.

"If that happens, I'll be twisting our man to the right to clear that area," Jim continued as Oak Bluffs Drive appeared on their right side.

"If there's another shooter in the room, I'll have my gun ready to take him out," Jim added as they turned right.

"Take it slow," Jim said as he looked at the nice houses in the neighborhood. "We don't want to draw any attention.

"There it is," he said, pointing to the large house at the end of the road. "Looks like you can just pull up to those garage doors on the right side, and we'll be out of sight until we reach the door."

"He'll probably have security cameras," Fabio mentioned as he eased around the circle leading to the house.

"Probably? Hell, definitely," Jim said as they pulled into the concrete driveway leading to the four-car garage. "Little late to worry about that now. Especially after the film festival we probably starred in a few minutes ago."

"Did you bring any Flex Cuffs?" Jim asked before opening his door.

"Wouldn't leave home without them," Fabio said, reaching into the console and pulling out several.

"Keep them handy," Jim said as he got out of the car.

Quietly shutting their doors, they stayed close to the brick and stone front of the house, and Fabio made a last check of his shotgun.

Standing in front of the door, Jim knocked loudly with the butt of his pistol. Waiting mere seconds, he knocked again more aggressively. Seeing someone approaching through the frosted pane of glass in the door, he put his hand behind his back and waited for the door to be opened.

Chapter 37

"Good morning, Mister Li," Jim said with a barely perceptible nod. "How are you today?"

"Who are you and what are you doing here?" Ming answered.

Jim quickly pulled his hand from behind his back and shoved the pistol beneath Ming's chin, saying, "I've come to tell you that the lady who said she was coming here isn't coming.

"Matter of fact, she isn't going anywhere … except maybe the morgue," Jim replied, pushing him back into the house. "So, you got your wish. But it came with a price. And that price is me."

As they stepped into the room, Jim saw two large men wearing suits, one on each side of the room. "And, Mr. Li, you should probably tell your pet gorillas not to make a move that might upset the man standing behind me. He's not a very pleasant fellow and takes serious offence at rudeness or aggression.

"So, I'd strongly advise you to make sure they remain nice and quiet until we finish our discussion," Jim added, pushing his pistol a little harder into Li's throat.

"And I'm a fairly nervous person myself, so try not to make any sudden moves. That would be the wrong thing to do right now with this pistol pointed up toward your brain. Do we understand each other?"

"You can't come in here without a warrant," Li complained. "If you don't have a warrant, I suggest you leave. And take your unpleasant man with you."

"I'm so sorry, my dog ate the warrant," Jim told him, keeping the pressure against his throat. "And I forgot to get another one. So, let's forget that warrant thing and just have a nice, pleasant conversation. It's just a piece of paper anyway. Besides, I brought this gun. And that's what you need to be concerned about right now."

Continuing to push Ming further into the room, Jim asked, "Are there any other people in the house? Is it just you, Ching, and Chang?"

"It is none of your business who is in my home," Ming replied. "And, I'll have you know that I have diplomatic immunity. And that includes my residence."

"Damn, that's nice to know, Mr. Li. Or is it Ambassador Li?" Jim answered. "Well, I'm not a diplomat. And the man behind me is very undiplomatic. So shall we cut the shit, and you just tell me what I want to know."

"Gun, Jim," Fabio said as he stepped left and lowered the Bullpup, firing at the man on the left as Jim spun Ming a quarter turn to the right.

Fabio's gun roared as the bullet from the man on the left zipped past where he had been standing less than a second ago.

The fifteen eight-millimeter pellets from Fabio's gun hit the man just below his belt, and the impact threw him against the wall behind him.

Simultaneously, Jim swung his pistol to the right, centered on the man reaching for his weapon.

"Do it, and you're dead," Jim said, looking at the man. "Sneeze … and you're dead. Cough … and you're dead."

Pausing to see if he made a move, Jim then said, "Slowly. Very, very, slowly, put your hands over your head. Keep them there until told what to do."

"Search him," Jim told Fabio, who had now aimed his Bullpup at the guy.

"Turn and face the wall," Fabio ordered him as he got three feet away.

As the man turned, Fabio nudged him in the back with the shotgun and said, "Okay, move to the wall and stop one foot from it."

When he was where Fabio wanted him, he continued, "Now, lean forward and put your hands on the wall, just above your head."

As the man did what he had been told to do, Fabio then said, "Now, spread your feet as wide as your shoulders and stay there until I finish searching you."

Jim told Ming, "Okay, slowly turn to face your man, put your hands over your head, and lace your fingers together."

As Fabio searched his man, Ming said, "I don't know who you are or what you want, but I'll do anything you ask as long as you don't hurt either of us."

"I don't believe you," Jim said as Fabio removed a nine-millimeter from the man's shoulder holster and another small caliber pistol from an ankle holster.

"I've already asked you if there were any other people in the house, and you refused to answer," Jim said. "Why should I trust you to do what I ask now?"

"Okay, Mr. Ching or Chang, whatever your name is, keep your hands on the wall and drop to your knees," Fabio said, holding the man's pistol to his head. "And do it slowly. Very slowly."

Once on his knees, Fabio told him, "Now, lean forward and put your head on the wall."

Once done, Fabio continued, "Now, put your hands behind your back and keep them there. I hope I don't have to remind you; I get very offended if you do anything that might surprise me."

As the man put his hands behind his back, Fabio took a pair of Flex Cuffs and pulled them extremely snug, saying, "If you move. If you attempt to get up. I'll cut you in half like I did your dead friend. Do I have your promise to be nice?"

"Yes," the man said without turning his head.

"Good," Fabio said, stepping back.

"Okay, Mr. Li," Jim said. "Let's say we start over. I'll forget the first question and ask another. Are there any other people in this house?"

"No," Ming answered.

"That's much better," Jim said, pushing him over to the wall beside his bodyguard. "But you know, I don't really believe you. So, I'll ask you to wait until I take a look for myself.

"Please kneel down by your friend, put your head on the wall, and your hands behind your back," Jim continued.

When Ming did as he was ordered, Jim told Fabio, "Cuff him and take a look around the house while Mr. Li and I discuss a little personal business."

As Fabio headed down the hall leading to the rest of the house, Jim quietly asked Ming, "Are you familiar with Lingchi?"

"You know … it takes many forms," Jim whispered. "Sometimes it's what you do to a person where it can't be seen. Like making a twelve-year-old little girl perform sexual acts on stranger after stranger after stranger until she's of no further use to you."

Chapter 38

A few minutes later, Fabio returned saying, "I think there are some people in one of the rooms in the back part of the house, but the door is locked.

I thought I heard some noise and listened at the door," he continued. "Sounded like at least three or four people whispering. What do you want me to do now?"

"Where's the key, Ming?" Jim asked, putting his pistol against the back of his head.

"I don't have it," Ming answered.

"I didn't ask if you had it," Jim said, leaning close to his ear. "You've got to learn to pay attention. I asked where the key was, not if you had it."

Pausing for a second, Jim repeated, "Where's the key, Ming? Last chance before I pull the trigger, blowing your piece of mind on the wall your head is resting on. Sorry, I guess I should say *a piece of your mind*. Then we'll ask your friend who's kneeling beside you, where it is."

"He doesn't have it," the bodyguard said. "The guy you shot has the keys."

"Well, well, well," Jim said, pulling his pistol from the back of Ming's head. "Finally, someone who understands the civil process of a pleasant conversation. Fabio, please go check and see if our new friend is being honest."

As Fabio approached the body lying in a pool of blood missing most of his stomach area, he said, "Damn, I should have shot him in the face. If the key is in his pocket, as I suspect, I've got to first find the pockets. Looks like they are somewhat mingled with what's left of the man's lower intestines."

"It's a dirty job, but somebody has to do it," Jim said. "I don't suppose you brought any neoprene gloves."

"I didn't think I'd be giving a rectal exam," Fabio said as he dug through the mess.

"Bingo," Fabio said, standing and holding up a key chain with several keys attached. "Shall I go see what lays behind door number one?"

"Please do," Jim answered. "And check the other rooms on your way back."

"I'll be back in a second," Fabio said, heading back down the hall.

A couple of minutes later, Fabio called saying, "You've got to come see this, Jim."

Jim leaned over to Ming and told him, "If you or your friend even move, I'll have my man gut you like he did your other friend. I hope I've made myself clear these last few minutes; I don't like it when someone disappoints me."

Rising, Jim said, "Come watch our host and his manservant while I take a look. I just don't trust either one of them for some reason."

"Damn, Jim," Fabio said when he got to him. "I don't understand how people can do the shit they do to other people. Especially kids. Some sick sons-of-bitches. Damn."

"Just don't shoot these guys," Jim said, turning to the hall. "Or at least until we have no further use for them. Which room?"

"Next to the last one on the left," Fabio said, looking at Ming and shaking his head. "It's the only open door."

Jim headed down the hall and stood staring at a room with bunkbeds lining the walls and several kids dressed in faded green gowns. The stench was almost overpowering as Jim looked at the frightened children.

Stepping backwards into the hall and continuing to look with disbelief at the scene in front of him, Jim pulled out his phone and asked, "Minnie, is the General still there?"

As Gene answered, Jim said, "Sir, we need to get a bus and a couple of ambulances over here as soon as Tammie can arrange it."

"What's the problem?" Gene asked, signaling for Tammie to join him.

"There are at least a dozen or so kids here who don't look like they've had a bath in a week, probably fairly recently arrived, and I'll bet don't understand a single word of English.

"You better have a couple of the Mandarin speakers come as well," Jim continued. "I don't think I've ever come across such a display of inhumanity and I've seen a lot of sick shit."

"I'll get the buses and ambulances on the way," Gene said, looking at Tammie and nodding. "Anything else?"

"Yeah," Jim answered. "I'll need maid service to clean up a mess Fabio made, and a van to escort at least two suspected illegals. Oh, by the way, one of them, a Mr. Ming Li, says he has diplomatic immunity. You might want to check that out.

"I'm sure Debbie's facial recognition program will have him if he is an ambassador or something," Jim continued. "Both Fabio and I gave her a good look at all the adults we've encountered so far.

"I'll let you know what else we find after we make a thorough search of this place," Jim said quietly, shutting the door and locking it. "I just hope we don't find another room like this one."

Chapter 39

Jim walked back to the front of the house, stopped beside Ming, and kicked him in the stomach as hard as he could saying, "You worthless piece of dog shit. I ought to blow your worthless head off right here and now."

Ming landed on his side next to his bodyguard, trying to catch his breath as Jim drew his foot back to kick him again when his phone rang. "Yeah," he answered, looking down at Ming.

"Debbie identified Ming," Gene told him. "The other two guys as well."

"Illegals?" Jim asked, still looking down at Ming.

"No. The two guards are here on work visas," Gene answered. "And Mr. Ming Li is here on a diplomatic visa. Some kind of cultural attaché."

"Well, he's about to become some kind of dead attaché," Jim said. "I'm going to put a bullet in his child molesting brain."

"Don't do that, Jim," Gene ordered. "We need him."

Letting Jim have a moment to calm down, he continued, "That man has some pretty high connections with the

Chinese Communist Party. He's also a member of the Triad."

"So?" Jim asked. "Does that give him a free pass on smuggling kids into our country and then using them in a child sex ring? I'd say his ticket home should be canceled and punched right here on the floor where he's lying."

"Just listen to me, Jim," Gene told him. "The man has value to us. By us, I mean our country."

"What sort of value do we have with a child predator?" Jim asked, shaking his head.

"He'll be used in a prisoner swap for a couple of English reporters," Gene explained. "We've been negotiating with the Chinese for a couple of years to get the reporters back to England.

"Up until now, we really haven't had much leverage," he continued. "This guy is most likely the best chance we'll ever have to get the English guys home."

"And he just walks away after everything he has done to these kids?" Jim asked.

"Of course not," Gene assured him. "He'll be tried here in the US for everything we can legally charge him with, including a few extras we might think up. He'll probably be spending the next couple of years in a special prison for sex offenders until the US, the English, and the Chinese decide how to make the trade."

"Then he just takes a first-class flight back home to enjoy fish heads and rice with his wife, and molesting his nieces," Jim said. "What about these kids? Who is looking out for them? This just isn't justice."

"You have to remember, Jim," Gene told him. "We were contracted to help remove illegal immigrants. That's what we're going to do. That's all we're going to do.

"This guy isn't an illegal immigrant," he continued. "So, we are under no obligation to deal with him, regardless of how we feel. And we certainly don't have any authority to do anything with him.

"Just do your job," Gene concluded. "Let the Feds take care of this. That's what they are for. That's their job.

"Search the rest of the house and then turn it all over to the folks I'm sending over there as soon as I get off this phone," Gene told him.

"And Jim. I hope the man doesn't fall down a flight of stairs more than once or twice," Gene finished. "I'll give you a call when I make arrangements for Mr. Li to be taken into custody."

"Understand, sir," Jim said. "And I'll call back when we finish searching the house."

Jim hung up, looked at Fabio, and said, "Why does the law seem to always get in our way instead of helping us get rid of the trash?"

"I blame it on mushrooms," Fabio answered.

"Mushrooms? How the hell do you blame it on mushrooms?" Jim asked as he turned from where Li was grimacing from the pain of Jim's last kick.

"Remember all of those hallucinatory mushrooms the flower children were taking back in the mid-sixties?" Fabio asked as he tried to pull Jim away from the cuffed men on the floor.

"My theory is that they all became legal aid attorneys and revised every law to cover their tracks for things they may or may not have done, but were too stoned to remember," Fabio explained as he pushed Jim toward the hall leading back into the house. "Now, we need to get a look at the rest of the house before the buses and ambulances get here.”

“I know,” Jim finally said as they came to the door opposite the room where the kids were. “I’m hoping that at the end, people like Ming are sent to Dante’s tenth circle of hell.”

“If I’m not mistaken,” Fabio said as he unlocked the door. “There are only nine circles.”

“There’s got to be a worse place for people who would allow the abuse of children the way that man has. Hell, allow is not strong enough. Promulgate is more like it,” Jim said as the door swung open. “Profiting from the sexual abuse of children deserves the deepest bowels of hell.”

“Holy shit,” Fabio said as they stepped into the room. “Look at all this crap.”

Looking around, Jim saw containers of a white powdery substance and stacks of baggies apparently filled with the powder.

A table where numerous forms of pills and capsules sat beside an automatic capsule counter with dozens of vials, scales, and hundreds of boxes of new baggies.

“I think we’ve found the other side of the Triad coin,” Jim said, looking around. “Sex and drugs. Add this to the marijuana trade, and you’ve got a major money-making machine.”

“Speaking of money,” Fabio interjected. “What do you suppose is in here?”

Looking toward the back wall where Fabio had slid a closet door open, he saw a built-in safe measuring approximately ten feet across with double doors, six feet high, and about four feet deep.

"Well, I guess I'd better make another call to Gene," Jim said, removing his phone and stepping into the hall. "I was told that we're only here to manage the illegal immigrants. The children appear to be the only thing we're to be concerned about."

As the phone was answered, Jim said, "We have another issue to pass off to the Feds, sir."

Waiting for Gene to respond, he continued, "We discovered what appears to be a major drug distribution center. Since I'm positive Mr. Li was involved with the prostitution, I mean the massage parlors, I'll bet my last penny we will find the links between here and the customers in those parlors.

"Maybe not just the massage parlors either, possibly the nail and hair salons," Jim continued, looking at the massive operation contained in such a small space.

"Anyway, I plan on leaving Fabio here to guard the place until you send the DEA people to deal with this, buses for the kids, and the ambulances for whoever needs medical attention," Jim told him. "And by the way, I'm sorry to have to tell you that one of the diplomats took another nasty tumble down the flight of stairs."

"Very strange to have stairs in a one-story house, isn't it, Jim?" Gene asked. "The FBI is already on their way for the two adults. Ambulances and buses are being worked on and should be coming within half an hour. I'll call the DEA and get them headed your way as soon as I get off the phone. Is there anything else?"

"One more," Jim answered. "Can you send a couple of ladies out here right now to take care of these kids? If we can just get them out of that room, maybe let them take a bath, clean them up. Find some clean clothes. I don't know. Just things kids who've been seriously traumatized need.

The rest of this shit, including the two adults, who will remain cuffed until your people get here, can wait. But those kids need immediate attention and neither Fabio nor I are the right people.

"They need to see a kindly woman's face," Jim advised. "Until you get a couple of ladies here, I'm going to leave their room locked. I have no idea what they would do if released right now."

"I'll get a couple of our Mandarin speakers there as soon as I can grab them," Gene promised. "I know Butch had a couple that were at the apartment where the lady kidnapped Jewell. I'll get them and have a Suburban running their red and blue lights, get them there as soon as possible.

"By the way, we haven't talked about her, and Butch hasn't heard, what's the situation with Jewell?" Gene asked.

"I'm not positive," Jim answered. "The EMT seemed to think she'd be okay, but to be honest, I haven't had time to check on her. I'll make a quick stop by the hospital where they took her on my way back and see how she's doing."

"Call me when you find out," Gene said. "And let me know if we need any other folks out there to wrap up that little diversion."

Chapter 40

Jim walked back into the room and asked, "How much of my conversation with Gene did you hear?"

"Most," Fabio said. "The only part that I really paid attention to was the part about me staying here until the cavalry gets here."

"I understand," Jim said. "But I need to get back and start doing what I was brought out here to do. I'm sorry, but we can't just drive off with those kids locked in their room and those two pieces of shit left unguarded.

"Not to mention, Mr. Li is some sort of important man to our government," Jim told him. "So, without many options, actually only one, I have to leave you here until the buses arrive to get the kids.

"You've got your pistol and the Bullpup, so you shouldn't have to worry about anything," Jim continued. "And I'm guessing it won't take Black Water too long to tell both the DEA and someone from the State Department about our special guest.

"I guess that's my way of saying I've got more to worry about than this right now," Jim said, putting his hand on

Fabio's shoulder. "I know you were only supposed to be my driver, but you know you have to be flexible in these situations."

"I understand," Fabio said, nodding. "I guess I feel demoted to being a babysitter instead of doing what I was hired to do."

"When you were told you would be my driver, what did you think you'd be doing?" Jim asked.

"Taking you from the motel somewhere once or twice for something," Fabio admitted. "Maybe running to get something for you while you were busy. Hell, I don't know."

"Did you ever imagine you'd be cuffing a child sex trafficking pervert, blowing the stomach section out of the pervert's bodyguard, finding a major league drug operation, and helping rescue a bunch of kids from a most horrible life?" Jim asked. "And let's not forget high-speed chases.

"All of that before lunch," Jim told him. "Look at this short break for what it is. A chance to sit back and think of all the stories you'll be able to tell your friends and neighbors at the next neighborhood bar-b-que."

"Not to differ, sir," Fabio said, following Jim to the front of the house. "I'm pretty certain the contract I signed with Black Water specifically addresses that. And I'm pretty sure I didn't see any exceptions for neighborhood bar-b-ques."

"You could be right," Jim said, walking over to where Ming lay. "Maybe you can write a fictional story, and the movie industry will hire some little twerp to play your part. Just think of how you'll feel being on your two-hundred-foot yacht sailing down in the Bahamas, hosting bikini contests sponsored by some suntan lotion company."

"You know, I'm beginning to think Mike was right when he called you an asshole," Fabio said as Jim grabbed the Flex Cuffs behind Ming's back and pulled his arms almost perpendicular to his shoulders.

"Just checking to make sure the cuffs were secure," Jim said as Ming screamed. "Can't risk having him get free before the folks from the State Department get here to save his sorry diplomatic ass."

Letting go of the cuffs, Jim said, "Now, Fabio. You shouldn't believe everything Mike Knox says. He's as big a liar as Punxsutawney Phil. And that rodent is one of the most prevaricators of the truth ever known."

"Not to tell you anything you probably already know, but there's a chance some unpleasant people could arrive before the Feds get here," Jim said as they headed for the door. "Maybe you can locate where the security cameras are being monitored and watch for any unusual activity in the neighborhood."

"How do I get back to the hotel?" Fabio asked as Jim walked to the car.

"Gene's sending a couple of ladies to take care of the kids for now," Jim answered. "They should get here within half an hour. Have them bring you back when they come back.

"If you need anything, give me a call," Jim said, getting into the car. "With any luck, I'll be standing around with a glass of tea watching the green and red dots chase each other around the big screen. So, you'd probably be bored to death sitting around watching me watching them."

"Somehow, I don't see how you're going to manage to stay out of the field," Fabio replied. "I've been paying attention, and there are so many moving parts; there's always going to be a chance to get back out there. And I'm pretty sure that's where you'd rather be."

"Could be right," Jim said, starting the car. "Anyway, I'm going to stop in and check on Jewell before I get back to the hotel. With some real luck, you'll beat me back there."

As Jim was pulling onto Oak Bluffs Drive, he called Minnie and said, "Hey, Minnie. Could you please give me directions back to the hospital where they took Jewell? Somewhere around where we put her and the kids in the ambulances."

"I know the place," Minnie told him. "Health North. Just north of 116th on the west side of Meridian. Aren't you coming back here? Mike has been looking for you for over half an hour."

"I'm going to make a quick stop and check on Jewell," Jim said, turning left on Maple Ridge. "Shouldn't take but a few minutes. I just want to see if she's come to and ask the doctor how she's doing."

"What do you want me to tell Mike?" Minnie asked.

"Nothing," Jim said. "But you need to get Fabio on your open monitoring thing. He's guarding a couple of high-value targets and a bunch of scared kids. He needs to hear from someone to assure him he's not alone.

"Just do it as if you need some information," Jim advised as he headed south on Oak Road. "And keep him informed as to when the ladies Gene is sending to take care of the kids will get there."

"I almost forgot," Minnie said. "They are on their way. Matter of fact, you could pass them on your way back here. They're about twelve miles south of the hospital where you're heading. You're about five miles north of there, but those guys are probably going twice as fast as you."

"Guess I just stay on Meridian now until I get to the hospital, is that right?" Jim asked as he turned south.

"Yes, sir," Minnie replied. "Do you still need directions?"

"Nope," Jim said. "I sort of remember where I am now. Thanks."

"Oh, one other thing," Minnie said. "Debbie just called a couple of minutes ago and said there have been a couple of calls to Mr. Li's phone."

"Did they say anything?" Jim asked, seeing the hospital just ahead.

"Debbie told Amanda the callers just said hello a couple of times and then hung up when nobody answered," she answered.

"Okay, let me know if anyone else uses that phone," Jim said, pulling into the hospital parking lot.

Chapter 41

When Jim entered the hospital and got directions to the room where Jewell was being watched, he headed down the hallway and finally found the room. Stepping in, he saw Jewell sitting up, but apparently asleep.

Walking up beside her, he noticed the numerous wires connecting her to a machine and monitor, which showed several lines moving across it. Just as he was about to turn around and try to find the doctor who was treating her, Jewell opened her eyes and said, "Hey, Jim. What are you doing here?"

"Just came to check on you," he answered, stepping up to the bedside. "How are you?"

"Hell of a headache," she answered. "How are the kids?"

"I haven't heard," he told her. "The last I heard was when the ambulance picked them up.

"The EMT said they appeared to be all right," Jim told her. "But that's sort of out of my hands for now."

"And the woman who took us?" she asked.

"Dead, I believe," Jim answered. "The EMT who drove the ambulance that brought you here said he couldn't find a pulse, but he couldn't make the call. I'll check with the doctors before I leave.

"My main concern is how you're doing," he told her. "My only interest in the woman is knowing if she is dead or we need to send someone to make sure she doesn't leave here unescorted."

"I'll be fine," she replied. "I'm so sorry I let this happen."

"You didn't let it happen," Jim assured her. "It was just one of those things that happen. Not your fault."

"My first time back in the field, and this happens," she said, turning her head from Jim. "Guess I wasn't as ready as I thought. Think Gene will ever let me have another assignment?"

"I'd bet on it," Jim replied. "One incident, which wasn't due to anything you did, doesn't erase all the operations where you performed perfectly. Now's not the time to start worrying about tomorrow.

"One good thing that came out of this, you guys led us to a major part of the Triad here in Indianapolis," Jim told her. "If you guys hadn't been trying to run, we'd have never known about it."

"You'd have found it without me," Jewell argued. "She'd have made a break for it either way."

"I doubt if she'd had the chance without taking you hostage to get to the car," Jim reminded her. "So, in a way, you could argue that we'd never have known where she was going and never found the guy if she hadn't had you."

"That's a bullshit argument," Jewell said with a slight shake of her head. "You don't have to sugarcoat the fact that

I screwed up when I was supposed to be searching her for weapons. The fact is, I screwed up."

"Did you know she had a knife hidden somewhere when you started the search?" Jim asked. "No. How could you have known? You hadn't even begun to search her. The lady had the knife at your throat before you put a hand on her.

"That's not the first time one of us has had a weapon pulled before we begin the search," Jim assured her. "And I'm sure it won't be the last. Hell, how many prison guards get shanked just as they're about to search a prisoner? So, just put that out of your mind."

"I appreciate what you're saying," Jewell told him. "But it was still my fault. I'm not sure if I ever want to return to the field after this. Not only did I screw up, I put those kids' lives in danger. I'd never forgive myself if anything had happened to them."

"The entire thing wasn't supposed to happen the way it did," Jim explained. "The pit maneuver was supposed to just slide the car to a stop.

"The rollover was due to the SUV being top-heavy," he continued. "We certainly didn't intend for anyone to get hurt.

"Anyway, stop worrying about tomorrow," he finished. "You just do whatever the doctors say, and let's let the future take care of the future. Sorry, I've got to run, but we're still trying to find the rest of those kids and get them out of the hell they've been living in for so many years.

"I think this is the worst assignment I've ever been given. Most of the people we're looking for, the children, didn't have a choice in being here. It's the ones who brought them here and forced them into the life they're living are who I want. I'll be checking on you, Miss Jewell. And I'll put in a good word with Gene."

Chapter 42

As soon as Jim got back to the car, he called Mike and said, "Sorry I've been out of touch for so long. What's happening with the TDAs?"

"We're just about to hit the apartments," Mike answered. "I've been watching all the moving pieces and decided to wait until Jon wrapped things up at the motel.

"That way, I could use his men," Mike explained. "Without knowing if the phones are still being used, being used as decoys to lure my men into an ambush, or just plain ass left behind, I'm a little leery of spreading them out across so many buildings.

"That little screwup at the motel taught me that what looks like a canary may be a vulture," he continued. "So, now my plan is to hit the three apartment buildings one at a time.

"I want overwhelming forces," Mike said. "I just don't trust the numbers I've seen. Who's to say another forty or fifty TDA members didn't move into those same apartments? You said they were supposed to have gone to Chicago. Did they?

"Do you have any proof they did, or did they toss their phones, and we only found the ones who couldn't be accommodated in the apartments?" Mike asked.

Pausing for Jim to answer, he continued, "I don't want to send my men into a suburban warfare scenario unless I can ensure we aren't outmanned. If you don't agree, you can replace me."

"I think you have a valid point," Jim said. "I know it will take a couple of more hours to wrap it up, but I think that's time well spent to ensure the safety of our people.

"Part of the reason for striking every location at once was to prevent any of them from learning about any raids at other locations and abandoning theirs," Jim continued. "Well, the element of surprise is no longer a factor.

"After everything that came out on every news outlet in the country regarding the motel operation this morning, we can kiss surprise goodbye," Jim surmised. "Now, more than ever, we've got to be prepared for the TDA folks to know we're coming because they sure as hell are prepared for us now. I agree completely with your strategy, Mike.

"Now that you're consolidating your teams, who's going to lead?" Jim asked, heading south on Meridian.

"Jon," Mike answered. "I like the way he handled the motel, and I know him better than the three guys who were going to be handling the three apartments. I'll leave it up to him, but he can let those guys keep their teams, just spread them across the different floors or hold some in reserve in case some of the TDA folks come strolling across the parking lot.

"Since we're getting into the beginning of rush hour for the morning commuters, I'd rather have those apartments locked down and keep our targets there than having to chase them across the city," Mike concluded.

"I agree," Jim said, noticing an incoming call. "Let me know when you plan to make your raid on the first apartment. What's your guess as to how long it will take to clear it out and have the buses on their way?"

"I hate to say it, but I'm estimating a half hour to an hour," Mike answered. "I'm going to treat each apartment in each building as if it contained the TDA. I don't want to be hitting apartment 207 and have someone from 209 step into the hall and shoot one of my guys in the back.

"If you'll find me another hundred men, I can cut that time in half," Mike told him. "Considering all three of the apartment buildings, my guess is noon to finish all of it."

"I don't have another hundred men, and I'd rather take three extra hours than lose a single man," Jim replied. "I think we oversimplified this because we had such superior, infallible intelligence. Well, that dog bit us in the butt, didn't it?

"I haven't seen how the other cities are going, but I'm guessing they're having the same issues we're having," Jim continued as he crossed 465. "I've got to take another call, Mike. Let me know when you're ready to hit the first apartment. And tell Jon I said, 'good job'."

"Butch, how's it going?" Jim asked as he answered the phone.

"Not as good as I had planned,' Butch told him. "After the incident at the apartment when Jewell was kidnapped, I heard that the other apartment building must have gotten word of the raid, and a couple of the ladies grabbed their girls and drove out of the parking lot.

"I didn't have any way to chase them except the buses, so I decided to let them run and see if we couldn't locate them later," Butch explained. "I figured if we had our eyes on them in their apartments, we could surely follow them on the road."

"That's pretty much true," Jim said. "The only fly in the ointment will be if they didn't take their phones. Unless you happened to get the license plates for the cars they were driving."

"Nope, we didn't," Butch said. "Our guys were mostly entering the buildings when the ladies made their break for it. Two of our team members were monitoring a couple of emergency exits, but the ladies entered through a rear door in one of the apartments.

"We didn't even realize who they were until we saw the girls with the ladies," Butch explained. "The man we had watching the emergency exits didn't notice them until they were already at the parking lot.

"He almost took a shot but was afraid of hitting one of the kids and alerting the building," Butch continued. "After the incident at the motel with everyone being alerted by a gunshot, we didn't want the entire apartment complex emptying out."

"You're probably right," Jim told him. "I'll get back to you after I get back to the hotel, where I may be able to see what's happening with the two runaways."

"I thought you were at the hotel," Butch replied. "What happened?"

"Long story," Jim answered. "We'll get into everything later this evening when everything is concluded. I hope. Let me know if there are any more issues getting the remainder of the kids at that apartment."

Chapter 43

Jim was just about to cross beneath 465 when he called Fabio, saying, "Fabio, boy, I have some good news for you."

"I can leave now, and you're taking me to lunch?" Fabio answered sarcastically.

"That's pretty close," Jim replied. "Butch had a couple of ladies escape from the other apartment building. I can only envision two places they would go. One is to the parlors or salons where they work, and the other is right where the lady we chased was going.

"If I had to choose one option, I would guess it would be to come there," Jim continued. "But I do have a way to cover both options. And you can already guess how I'll handle at least one of the places."

"You want me to sit here and wait for them?" Fabio replied.

"Only until I can rule it out," Jim told him. "I'm hoping they still have their phones, and we can trace them. I'm also thinking that if they are coming your way, they will call like the lady did before.

"So, you can go into the kitchen and make yourself a sandwich, grab a Dr. Pepper, and watch Hollywood Housewives on TV", Jim said. "Or, and I think this is a more productive option, you can take the keys and start a detailed search of the house while waiting for the DEA, FBI, ICE, or whoever gets there first.

"Also," Jim continued, "it would be a big help if you'd go get Ming Li's phone and make sure it's turned on. Just in case one of the two runaways do call him. I'd like to keep them on the phone as long as possible."

"You expect me to engage them in some conversation?" Fabio asked incredulously. "I'm from Cuba. I barely speak English."

"I don't expect you to talk to them," Jim explained. "Just let them say 'hello' a dozen times before they give up. Or you can rub the phone against your jeans, trying to simulate static.

"It doesn't matter either way," Jim finished. "We can track the phones if they have them with them, but if we know they are trying to contact Ming, it will probably mean that's where they want to go."

"I understand," Fabio replied. "What should I tell the DEA, FBI, or the WTFs when they get here?"

"Just what you know," Jim answered. "If you discover any more kids, drugs, or money, let me know because we may need to send an additional bus."

"I'm guessing you want me to give the DEA folks the money," Fabio said. "Isn't there something like a ten percent finder's fee or something like that?"

"Go for it," Jim told him as he crossed Kessler Boulevard. "But if that's your plan, you better find the keys to Ming's car and start loading it now, because I just heard the ladies and ambulances are pulling up now to get those little kids.

"I still would like for you to make another pass through the house," Jim told him. "Just to make sure there isn't some other room or cellar where any more kids are being kept. Or money. Give you something to do while waiting for those kids to be cleaned, or fed, or whatever the ladies coming to get them think needs to be done.

"Let me know when the Feds get there," Jim finished. "I've got to see if I can figure out where Butch's escapees are heading."

"Minnie, are you still on?" Jim asked.

"Of course," she answered.

"Please let me know if we're able to find and follow those two ladies who got away from the apartment Butch was going to raid.

"I'll check with Quantico," Minnie answered. "I think they were covering all of the phones from both the Triad and TDA for us."

"Let me know," Jim told her. "And get someone from our group to find them and keep track of where they are going. I'd like to get in front of them and apprehend them as soon as possible.

"Amber, are you still keeping track of me?" Jim asked as he crossed beneath I-65.

"Yes, sir," she answered. "Both you and Fabio."

"Good," he told her. "As you can see, I'm almost back at the hotel. I want you to keep an open phone line with Fabio and let him know if any of our Triad targets are headed his way.

"I don't expect anyone to come after me unless it's because I'm going to leave this car parked just outside the entrance," Jim told her. "But there's a chance someone from the Triad who doesn't know what has happened may try to contact Mr. Li and head out there when they don't get an answer.

"There's too much cash and drugs lying around out there to go unnoticed," Jim continued as he looked down at his gas gauge. "Not to mention, someone may be coming for those kids. I'm sure he didn't run a long-term day-care center.

"Anyway, I've got to pull over up here for some gas, and I should be back there in fifteen or twenty minutes," Jim told her, spotting a self-service gas station just ahead on his right.

Pulling into the station, Jim made one more quick call as he stopped beside one of the pumps. "Butch," Jim said. "See how many people you can send to the parlors and salons. That's one of the few places I think these folks will go.

"I've got Fabio watching the only other place, and I'd like for your people to make sure they don't go to where they worked," Jim continued. "Another reason is that I believe they were also dealing drugs from those locations.

"Have your guys do a thorough search for any hidden rooms, closets, anywhere drugs could be stored out of sight," Jim finished as he opened the door to get out. "I'll talk to you when I get back to the hotel."

Laying his phone on the console, he pulled out his wallet and took out the only credit card he owned. After swiping his card and selecting the appropriate octane for the car, he inserted the nozzle and let it begin filling the tank.

When the hose finally kicked off, signaling the tank was full, he replaced the gas cap and hose, then started to get the receipt. Noticing that it didn't come out, even though the Printing Receipt notice was showing, he decided to go inside and get a Dr. Pepper, then have the cashier print one for him.

Turning right upon entering, he saw the soft drink section against the far wall. Getting his drink and looking at the chips, he noticed a guy wearing a hoodie walking in and stopping in front of the cashier.

As that didn't seem too unusual, he finally found a bag of lime-flavored Takis and headed for the cashier. Almost there, the guy in the hoodie turned slightly to his right and looked directly at Jim. The first thing Jim noticed was the gun in his right hand.

Looking back at the man's face, Jim realized he was probably being scrutinized because he was still wearing the ICE jacket. Jim took a quick glance back at the gun and recognized the crown tattoo on the guy's hand. Knowing that was one of the identifying marks of the TDA, he wondered if Debbie's facial recognition program would be able to identify the man.

Thinking that Amber would notify him if that had happened, he suddenly realized that he had left his phone in the car and was out of contact with any of the group.

Holding the bag of Takis in his left hand and the Dr. Pepper in his right, Jim started squeezing the bag as hard as he could. As the material suddenly gave way with a loud pop, Jim dropped the can of Dr. Pepper and pulled his Glock as the target started to turn toward him.

Dropping to land on his right knee, Jim brought his pistol up and took a quick shot at the middle of the man's chest. When he knelt on his left foot and right knee, he took careful aim at his chest and fired another round.

As the man fell toward his right side, Jim kept his pistol trained on his head in case he wasn't dead. Once the guy hit the ground, Jim rose and kicked away the gun he had dropped. Satisfied he wasn't going to get up, Jim looked at the cashier's startled face and said, "I'll be right back."

Chapter 44

When he reached the car, he leaned in, grabbed the phone from the console, and said, "Amber, are you still there?"

"Of course," she replied. "That's my job. To be your guardian angel."

"Well, I guess it's my fault that I left my phone in the car, and you didn't have a chance to tell me a TDA guy followed me into the gas station," Jim told her as he headed back inside. "That's not important right now. Please tell Tammie that we need an ambulance sent to this location.

"And, if the General is handy, ask him if he can provide a city cop to come investigate a shooting," Jim continued as he walked to the cashier.

Stepping around the prone body, Jim handed the cashier a twenty-dollar bill and asked, "Where are the computers that monitor your security cameras?"

"I think they are in the back office, but I've never been back there," the cashier said. "If you need me to check, I'll call the assistant manager. He's about four blocks away at another one of our stations."

"Please do," Jim said as he took his change. "And please give me a receipt for pump number twelve."

As the cashier was handing him the receipt, Jim's phone rang. Looking at the caller ID, he answered, "Good Morning, Debbie. What can I do for you?"

"First, you can start carrying your damn phone everywhere," she scolded him. "Amber called and said she didn't know about the TDA guy, and I never got a look at him because your glasses won't transmit to your phone if your damn phone is more than fifty feet from the glasses.

"All the technology in the world is useless if you don't use it correctly," she continued. "Now, since you now have your phone and glasses within the correct distance, I can tell you that the man lying on the floor there is one of the TDA guys from Colorado.

"And, if you'd had your phone with you when you were in the store, we could have told you he was coming," she told him. "Now, I hope you've figured out that neither I nor Amber can do shit for you if you don't operate the equipment correctly."

"I understand," Jim said contritely. "I hope it doesn't happen again. Now, since you've identified this man as one of the Colorado escapees, can you tell me where he's living here in Indy?"

"Not exactly," Debbie answered. "We think he was in one of the apartments, but it seems that the phone he has with him is being used by several of the guys from Colorado. It has been in different apartments several times, and we've identified several different voices that are using it, so our assumption is that they have tossed most of their phones and are using just one to keep from being tracked."

"Sounds reasonable," Jim said as the cashier signaled to him. "Sorry about the screwup, Debbie. I'll definitely keep the phone in my pocket from now on, but I've got to take care of something right now."

"Fine," Debbie replied. "Just remember, the phone and glasses are a set. Just having your phone isn't the answer. Neither is having just the glasses. They have to be together. Otherwise, they can't work as designed."

"Got it," Jim told her. "I'll flog myself when I have time. Twenty lashes. Now, I've got to take care of a couple of things before I can chastise myself appropriately."

Turning to the cashier, he asked, "Is the assistant manager coming?"

"No, sir," he explained. "When I told him what happened, he told me to call the manager. He should be here in ten minutes."

"Fine," Jim told him. "I've already called 911 and was told the ambulance is on the way. I also heard that an Indy police car is coming. So, you don't have to worry about any of that.

"But you could have a problem with anyone coming into the store," Jim continued. "I suggest you ask your manager or assistant manager if it would be okay to put a closed sign on the door and lock it. I'll move my car to the front and leave my red and blue lights flashing."

Seeing the confused look on the clerk's face, Jim said, "Never mind. I'm going to move my car and stand outside to keep people away. They can continue to use the self-service pumps, but I can't allow anyone to contaminate the crime scene. I'll square it with your manager."

Seeing Gene's number flash on his phone, Jim turned away and headed back to his car, answering, "Sir, I'm afraid

I've created a slight problem, and I need some assistance to help keep this from being all over the news."

Listening to Gene asking about what had happened, Jim got into his car and moved it to the front of the gas station's doors.

"Well, sir, I shot a TDA, and the clerk definitely watched," Jim said as he turned on the flashing lights. "I hope I've stopped him from calling 911 by lying to him, saying I had already called.

"Now, the manager is on his way," Jim continued, getting out of the car and standing in front of the glass double doors. "I'll get the security tapes and try to convince him from getting further involved.

"But, if we don't get our ambulance here before anyone else calls one, we'll have a tough time getting the body away," Jim told him. "And we certainly don't want what you called an uncooperative city cop to be the first on the scene."

Listening for a moment, Jim explained, "I know I said I had called 911, but I didn't. However, I told that to the cashier. I don't know if he told the manager or assistant manager, or if they made a call themselves.

"All I know is we need to get our people here as quickly as possible, get the security camera tapes, and get the body," Jim continued. "I don't want to have to stay in front of this door trying to keep people out for any longer than it takes to get someone else over here."

Listening to Gene say help was on the way, Jim saw a car stop at one of the pumps, and a woman get out, walking toward him. 'Well, shit,' he thought to himself as he walked around to the driver's side of his car to intercept her. 'Guess it's time to put on my official face and see how persuasive I can be.'

Chapter 45

"Can I help you, ma'am?" Jim asked as the lady approached him.

"I just need to get some gas," she answered.

"I'm afraid the only way to get any gas right now is to use a credit card," Jim told her.

"I don't have a credit card," she told him. "I just need to get twenty dollars' worth, and I'll prepay, like I always do."

"The station is closed, at least inside where the registers are," Jim explained. "But I'll see if I can help you. If you give me the twenty dollars, I'll take it inside and have him turn on a pump. Which pump are you going to use?"

Looking over her shoulder, she said, "Number four. Twenty dollars of regular, please."

"Hang on just a second," Jim said, knocking on the door.

"Why is the station closed?" she asked, trying to look around Jim.

"There was a robbery attempt," Jim explained. "And someone was shot. I'm just helping until the police get here."

"Was anyone hurt?" she asked, still trying to get a look inside.

"Well, I think just the guy who was shot," Jim answered as the cashier came to the door.

"The lady would like twenty dollars' worth of eighty-seven octane on pump four," Jim said, handing him the twenty-dollar bill.

"The cashier says the pump is ready and you can get your receipt when you replace the nozzle," Jim told her as he heard the distant sound of sirens.

No sooner had the lady driven off than a car pulled up behind Jim's, and a man wearing polyester pants and a beige knit shirt with the station's logo over the left breast pocket got out.

"Who are you and what are you doing here?" the man asked.

"I'm with ICE," Jim answered. "I'm just trying to keep people out until the police get here. I didn't think you'd want people to come in and see a body in a pool of blood, and I know the police wouldn't want anyone to contaminate their crime scene."

"Well, I'm not just anyone, I'm the manager, and I need to get inside," the man said.

"Not a problem, sir," Jim said, stepping to the side. "I've been asked to have you make a copy of your security tapes, and I'll deliver them to the police as soon as they get here."

"Just exactly why should I give you the tapes?" the manager asked as the cashier unlocked the door.

"Because the man on the floor is a member of a gang we've been looking for and because it's a federal investigation," Jim told him, following him into the store.

"My cashier told my assistant that you were the man who shot that guy," the manager said, heading for his office in the rear. "If that's true, aren't you a suspect?"

"No, sir," Jim answered, following him. "I'm not a suspect. I am, as you said, the guy who shot him. Now, I need a copy of the tapes for our investigation, and we'll provide the police with a copy for theirs. By the way, does your security system store the data here or off-site?"

"It's transferred to the regional office," he answered, picking up the office phone. "They'll have to approve giving you a copy."

"Fine," Jim said, turning to step out of the office and removing his phone. As soon as it was answered, he said, "Debbie, can you get into a company's security camera system?"

"Of course," she answered. "What do you need?"

"This station, where I'm standing, sends its security cameras' data to a regional office," he explained. "I need you to get into their system and delete the records from this station for the last hour or so."

"Can you get me the phone number?" she asked. "That's probably the same line the computer uses to send the data to the regional office."

"No problem," Jim told her, stepping back into the office. "I'll just look at the phone and the equipment the cameras are connected to, and you should be able to see any numbers you need."

"I'm sorry," the manager said, hanging up the phone. "My boss says we need a subpoena before we can turn over any records."

"Did you explain that I've asked for the records and that I'm a federal agent?" Jim asked.

"Yes, sir, I did," he answered. "But, he says, corporate policy is that no one sees our private data unless they get a subpoena. Not even the police."

"Could I get the phone number for the corporate office?" Jim asked, looking at the phone and the box where all the cameras were connected.

"Sure," the manager said, handing Jim a card from a holder on the desk. "Our company always wants to provide the police or federal officials, any assistance possible."

"Thanks," Jim said, holding the card up in front of him. "We appreciate that, and I'm sure someone from my office will be contacting you as soon as possible with a subpoena. And I'm sorry this had to happen at your store."

"I understand, sir," the manager said, leading them back to the front. "My cashier told me the man you shot was robbing the store and had threatened him, saying he'd kill him if he didn't give him all the money in the cash register.

"As much as we try not to antagonize any person who wants to rob one of our stores, sometimes they get very disappointed by the little amount of cash we have on hand," the manager continued as they reached the front.

"Most of our sales are on credit cards, and the cashiers have been instructed to put any money over what their opening money is, that's used to make change, into the drop slot of the safe just below the cash register," he explained. "We've had people terribly hurt when they can't give the culprits what they think they should have."

"I understand, sir," Jim said as an ambulance pulled up front. "Now, since the ambulance is here and I'm sure the police are not far behind, I've got to get back on duty myself. I hope you have a better day than this one seems to be, and I'll make sure someone comes by with that subpoena."

Chapter 46

"Did you get everything you needed?" Jim asked as he left the office, calling Debbie.

"Got it," she answered. "The data from that station has conveniently disappeared from any records."

"Great," Jim said, getting to the doors as the EMTs were coming in. "I don't think I need to have my picture splashed across the nation."

"Guess you haven't seen the news lately," Debbie said. "I think you'll be surprised at where your picture has surfaced."

"I'll call you back," Jim said, holding up his hand in front of the first EMT signaling him to stop.

Putting his phone in his pocket, he asked, "Are you from ICE or the city?"

Hearing they were from ICE, or more correctly, the *Black Water division* of ICE, he said, "The deceased is a member of the TDA who came up from Colorado. You need to take the body to the location where all the members are being held pending removal by the appropriate federal agency."

Hearing they had already received those same orders; Jim thanked them and jumped into his car.

"Amber," he said as he headed back toward the hotel. "Anything happening with Fabio?"

"No, sir," she said. "He called a few minutes ago and said he had searched the entire house and there was nothing else. Now he's just waiting for someone to come get him,"

"Someone should be there soon," Jim said as his call waiting light came on. "Check with Tammie and see if she can let Fabio know when the buses and ambulances we sent to take care of the kids will arrive. I should be back at the hotel in fifteen minutes or so."

Pressing the disconnect button, Jim answered the call that had been waiting, "What can I do for you, Butch?"

"We went through the parlors and stuff, but there weren't any other people there," he answered. "What else do you want me to do?"

"Go back and tear those places apart," Jim answered. "We found a shit load of drugs and stuff at the place where your Chinese lady was heading, and I'd bet a dollar to a doughnut that's where the central supply was for those places where the girls worked.

"So, go into all of them and look for hidden compartments, false walls, access to the attic or an underground room," Jim said as he headed south on Meridian. "While you're looking around, see if you can find any customer list or anything. I'm sure not everyone using the nail salon is just going there for the quality of glitter they want on their nails."

"We'll see what we can find," Butch replied. "Have you given up on either of the two ladies going to where they used to work?"

"Not entirely," Jim answered. "So, have your guys park somewhere discreet and be ready to take the women and kids if they show up. I'll let you know when we get a good idea of where they are headed. I'm still betting on the same place as the first one."

"Anything else?" Butch asked.

"Not that I can think of," Jim answered, crossing I-65. "Let us know if you find anything hidden in those parlors. And be careful, just a tiny amount of some of that shit, like fentanyl, can end your days."

"Better to just have them find the stash and notify our folks here and let them send the DEA guys in to deal with the crap," Jim said seeing his personal phone silently showing a missed call. "I'll talk to you hopefully in a couple of hours when this is over."

Making the left turn on North Street to join Pennsylvania Street, Jim wondered who would be calling him on that phone instead of the company phone they were all using.

As he turned south on Pennsylvania, he thought about what Debbie might have meant about his picture surfacing. Trying to recall any contact with the public, he could only think of one.

As Jim walked into the conference room, he spotted Gene on the phone and looking at it. Waiting until he looked up, he asked, "Have you seen any pictures of me back at the place where we stopped the car from the apartment? Where we put Jewell and the kids in the ambulances."

"You mean these?" Gene answered, showing him the pictures he had been looking at.

Jim looked and saw a video of him carrying one of the kids to the ambulance. "I guess you could say that. Is that the only one?"

"Nope," Gene told him. "There's more. Apparently taken from some car that was there shortly after you guys blocked the road. This one was probably somewhere close to the front."

Showing Jim several more clips of the video, he continued, "I don't know if there are others, but this one has probably made nationwide coverage by now.

"The only thing I can be grateful for is that whoever took it was shooting through the windshield," he said. "I just hope that none of the other people there got anything else."

"That's bad enough," Jim replied. "I hope they didn't get a shot of Fabio shoving that guy to the ground."

Gene scrolled through his phone and said, "You mean like this one?"

"Oh, shit," Jim replied looking at the video of Fabio.

"I've already talked to the head of ICE and made sure he knew it was us, so they're coming up with a cover story," Gene told him. "What the hell happened?"

"Our pit maneuver went a little south," Jim said, shaking his head. "Then the guy Fabio shoved wouldn't get back to his car as we asked. Things escalated a little from there."

"Well, at least there's not a good shot of either of your faces," Gene said. "But the ICE on your jackets shows up plain as day."

"What do you suggest?" Jim asked, looking at Gene.

"Just get this thing over with as soon as possible," Gene answered. "And try not to make any other public appearances. And stress to Butch, Mike, and their people, that incidents such as this don't happen. At least not in public. The blowback against this entire operation is causing some concern to the folks that hired us."

"Understand," Jim replied. "I'll pass it on. But those damn iPhone cameras are just about everywhere today."

"You already knew that," Gene told him. "For God's sake, at least put some lipstick on the pig before someone videos you shooting one of these poor immigrants who's only here seeking asylum."

Chapter 47

As Jim accepted the criticism, he walked over to Minnie's desk and asked, "Any word from Butch regarding returning to the parlors?"

"Not yet," she answered, zooming in on the strip centers where the massage parlors and salons were located. "They should be arriving soon."

"Good," Jim told her, finally looking to see who had called his personal phone.

Walking away, he saw Marie's number and decided to call her back.

"Marie, I saw you called," he said when she answered.

"Yes, I did," she responded. "Where are you? Are you in Indianapolis?"

Jim hesitated and finally admitted, "Yes."

"What are you doing up there?" she asked. "Are you flying or is this one of those things you do with Gene?"

"I'm with Gene," he answered as he heard Minnie calling for him to come to her desk.

"Look, I can't talk right now," he continued. "But we can talk about this tomorrow when I get back."

"Fine," she said and hung up.

'Well, shit,' Jim thought as he walked back to Minnie's desk. 'I wonder how she knew I was here?'

"What's happening?" he asked, looking at the screen where numerous green dots appeared moving to the parlors.

"Some of Butch's guys just arrived at this parlor," she answered. "The other groups are still enroute. The last one should be there within fifteen or twenty minutes."

"Keep an eye out for them and let me know when they get there," Jim told her as he headed to Amanda's desk.

"Amanda, can you put the phones of the group at the parlor on Limestone Street on the speaker?" Jim asked, looking at the screen.

"Give me a second to get their numbers," she answered. "It would make it easier if I had a list of the phones for each group."

"Maybe you can start making a list," Jim replied as she began punching it in on her keyboard. "When you get time."

"Okay, I've got them coming on the speaker now," she said as Jim heard the obvious team leader saying, "Okay, let's do a last-minute weapons check and be prepared to bust in the door if it's locked and we don't get an answer when we knock."

A couple of minutes later, Jim heard him say, "Bust it in. I want two men with shields to go first and clear the room before the rest of you go in."

Listening as the team cleared the place room by room, the leader said, "Okay, now start tapping on the walls, check the ceilings for access, and look for any way to get into any room beneath the floors."

"I've got one back here," someone said. "It's in the storeroom. Behind a closet. Do you want me to go in?"

"Wait for backup," the leader directed. "I want two men before we enter any hidden areas."

A few minutes later, the guy who had found the hidden door said, "Holy shit! There's a ton of shit in here."

"What is it?" the leader asked.

"Looks like the inside of a pharmacy," came the answer. "There are dozens of shelves with hundreds of baggies and bottles."

"Get out of there," the leader directed. "We don't know what's in there, and I don't want anyone touching anything that could kill you."

Just then, another man shouted, "I've found a trap door in the floor of the office."

"Wait for someone to get there before you open it," the leader ordered.

Moments later, the guy said, "There are several young girls hiding down there. Hell, it looks like a dozen or so."

"Okay, let's wait to see what Butch wants us to do," the leader said. "Do you see any adults with them?"

"Yea, she's sitting on the floor at the back of the room," he answered.

"Secure the area and standby," Butch ordered as he heard what had been discovered.

Almost immediately, Jim's phone rang. Answering, Jim said, "I heard. Have your guys shut those doors and stand by. I'm sending someone from the DEA as soon as we can arrange it, along with a Mandarin speaker. Don't let anyone from that basement out. I'm also sending an ambulance to look at the girls. By the way, does your team have any Naloxone with them?"

"No, we don't have any," Butch replied. "Do you think we need it?"

"Hopefully not," Jim answered. "But I want anyone who went into the room with the drugs to be checked by the EMTs when they get there. Get them outside so they'll be the first ones they look at. And have them stay away from the rest of the team.

"Leave enough men to cover the doors where the drugs and kids are. And put someone out front to ensure no one gets inside before the ambulance arrives, or the DEA gets there. If the DEA gets there first, make sure they are aware of where the drugs are and that we'll be taking care of the kids.

"Then start looking at the other salons there," Jim directed. "Just don't let your guys go into any room that looks like it may have drugs. Open the door, look inside, and shut it. I don't want anyone else to be exposed to whatever's in that room. That's the DEA's job."

Stepping over to Tammie, he said, "Get an ambulance to the parlor on Limestone as fast as you can. And advise them there's a possibility of contact with Fentanyl by some of our guys.

"And several young girls who may need medical attention as well," he added. "There's no telling what those kids have been exposed to."

"They just arrived at the parlor on Rockville Road," Minnie yelled at Jim.

Chapter 48

Jim stood looking at the screen and then remembered he hadn't talked to Mike since he had said he was about to hit the apartments where the TDA members were staying.

Dialing Mike's number, he asked, "What's happening on your end?"

"We've cleared out one of the apartments," Mike told him. "It went relatively smoothly. We're expecting quite a bit of resistance from the other two, but I'm guessing it should be done in thirty minutes or so.

"Jon is doing a hell of a job," Mike continued. "I think our biggest problem is the number of weapons these guys have. Granted, we've got better firepower, but they are in a defensive position, and general wisdom is that it requires three times as many on the attacking side to be successful."

"You should have that many, if not more, don't you?" Jim asked, looking at the group of green dots.

"Yes," Mike answered. "But this isn't conventional fighting. They've had warnings about our approach, and our points of access are extremely limited. I sure wish I had some airpower right now. A flight of four F4s carrying five hundred pounders with daisy cutters would solve my problem really quick."

"And level a couple of apartments," Jim replied. "I guess you forgot the low-profile discussion. If you want to make the news, that would surely do it. By the way, we've located the missing TDA from Colorado."

"Really?" Mike asked. "Where did they find him?"

"At a gas station about six or seven miles north of the hotel where we're staying," Jim answered.

"What the hell was he doing way up there?" Mike asked.

"Trying to rob a gas station," Jim answered.

"Did you capture him?" Mike asked.

"More or less," Jim told him. "I shot him."

"I didn't think you were supposed to be out in the field," Mike replied. "What the shit happened?"

"You remember when I said I was busy?" Jim asked him. "I said I'd call you back, but things got sort of sideways, and I had to make a sudden run to find a lady who escaped Butch's people.

"We can discuss it when you get back to the hotel this afternoon," Jim told him. "Is there anything you need to wrap up those two apartments?"

"I believe I mentioned F4's," Mike responded. "But if that's out of the question, another hundred men."

"I may be able to get you some help tomorrow," Jim said, laughing. "But let me know how it's progressing."

Hanging up, Jim dialed Butch and said, "I've got a little bad news for you."

"What's that?" Butch answered.

"When you get through with the parlors and other businesses, I need you to go back to the farms and do a search of the houses," Jim answered. "Since there were so many drugs in the parlor operation, I'd bet there were more of them in the farms."

"You know that's going to take another hour or so to finish this," Butch told him.

"I understand, but I think those farmers had more than just a marijuana operation out there," Jim argued. "Since the Triad was running both the parlors and farms, it stands to reason they had similar operations."

"Okay," Butch replied. "I'll let you know when we finish the parlors. Maybe I can break a few men loose to head out there sooner than an hour. The businesses besides the parlors have been pretty much empty, so I can probably reduce the number of guys on that operation."

"No, I don't want to take folks away from that," Jim told him. "There shouldn't be anyone at the farms either, so just keep your guys working there. I don't want to miss anything, especially if there are any more girls being kept in those basements."

"Sounds good, Jim" Butch replied. "But I can reduce the number of men at each farm and cover them a lot faster than we did on the first shot. Should cut the time in half. Especially since the hothouses have been torched. About all that's left is the house where they lived and an outbuilding or two."

"Okay," Jim told him. "Give me a call when you're ready to head back out there. And just because the houses were empty when you left them, don't assume someone else hasn't gone there. We don't have any red dots out there right now, but there could be some Triad folks without phones.

"We haven't lost anyone yet," Jim concluded. "It's not worth the risk just to save a few minutes."

Jim walked back over to Amanda and asked, "Any news about any of the ladies and their kids who made a run for it?"

"Nothing yet," she answered. "I've been paying close attention for any calls going to Ming's phone, and nothing yet."

"I'm sort of surprised," Jim said. "That would have been my first option if I was running."

Jim stood looking at the screen and went back to Tammie's desk, asking, "Do we have any information regarding the DEA and ambulances we sent to the first parlor where the kids were found?"

"Hang on," she told him. "I can check on the ambulances, but I don't have contact with the DEA."

"I'll be right back," Jim said, turning to look for Gene.

Seeing him near the back of the room using his phone, Jim walked toward him and stopped a few feet away to give Gene some privacy.

As Gene finished his call and looked at Jim, he stepped over and asked, "Do you have any way of finding out when the DEA will get to the parlor where we first found the drugs?"

"I can make a call," Gene answered. "Is it important?"

"I believe so, sir," Jim answered. "You remember the two ladies who got away from the apartments? Well, Butch said some kids and a lady were at the first parlor they hit. If they don't turn up any more kids, I'm guessing that those people hiding at that first parlor are one of them."

"And the second batch?" Gene asked, dialing his phone.

"Still a possibility they are headed to Ming's place," Jim answered.

"How are we going to prove the lady at the parlor is one of the two?" Gene asked, waiting for his phone to be answered.

"I'm going to have Butch tell one of his men who are waiting there to open the trap door and get a good look at the lady," Jim answered. "Then Debbie can see if she matches one of the two we suspect her to be."

"Go ahead," Gene agreed, hanging up his phone. "Let me know as soon as you find out. I'd like to get all the moving pieces nailed down before DEA folks start showing up and drawing even more attention."

Chapter 49

"Butch," Jim said when his call was answered. "I've got another request."

"Now what?" Butch asked.

"I need you to get one of the guys you left at the parlor to go take a look at the lady hiding with the kids," Jim answered.

"I bet you're thinking she's one of the runners from the apartments," Butch replied.

"Correct," Jim told him. "This shouldn't take long and won't change your people's locations. If you can get her out of her hole, that would be great. I'd prefer if there were some lights down there and your man could get a good look at her face, then Debbie can let us know if she's one of the missing ladies."

"I'll give them a call," Butch said. "I'll get back to you when I hear."

Walking over to Amanda, Jim asked, "Can you put Butch on the speaker right now?"

Amanda punched a couple of keys, and they heard Butch talking to one of the guys who had been left to guard the parlor.

After what seemed like an eternity, they finally heard the man at the parlor say, "I found a light switch and I got her to stand up, so I had a good view. What do you want me to do now?"

"Leave the light on and shut the door," Butch told him.

Hearing how it went, Jim dialed Debbie and waited for her to answer. When she did, he asked, "Did you get a chance to see if the picture from the parlor matches any of the ladies the Triad had working there?"

"Hang on," she answered. "I've got more pictures coming in than you could ever imagine. You're just one of several hundred people I'm monitoring. Give me a couple of minutes."

Seeing Butch's call coming in, Jim asked Amanda, "Can you just tell Butch we got the photo, and I'll let him know if it's her as soon as Debbie lets me know."

"You've got a match," Debbie said a couple of seconds later. "See how good this works when used properly?"

"Are you going to keep milking that dead cow forever?" Jim asked. "But thanks anyway."

Hanging up, Jim called Butch and said, "That's one of them. Now, all we need to do is find the other one."

"Any clue about where she is?" Butch asked.

"Not yet," Jim answered. "If you just happen to stumble across her at one of the other parlors, hiding like the one you found, let me know."

"We just got a call through to Ming's phone," Amanda said. "We're getting the location of the caller right now. We should have it in a minute or so."

As she was busy trying to pinpoint the location, Jim's phone rang. Looking at the caller ID, he answered, "Good timing, Fabio. Let me save you the trouble of telling me, you got a call on Ming's phone."

"How did you know?" Fabio asked.

"I have my moments," Jim answered. "We don't have a location yet. But I'm still guessing she's headed your way. Is the DEA there yet?"

"Nope," Fabio answered. "Not a sign of them or the ambulances."

"I'll see if I can get an update for you," Jim told him. "Just be watching for this lady and some kids. If the ambulances get there before this lady shows up, ask if they can wait until I know if she's coming your way."

"Maybe you should get a bus or two headed here," Fabio suggested. "If the kids don't need any medical attention, we can send them to the feds."

"I'll see if Tammie has any to spare," Jim told him. "Right now, I'm more concerned about getting our hands on the lady who called you."

Walking over to Tammie, Jim asked, "How many more buses do we have access to?"

Tammie checked her computer and answered, "I still have a couple. What do you need?"

"Not sure yet," Jim answered. "But you need to make sure the drivers are there and ready to roll within the next few minutes."

"What's the latest on the Triad operation?" Gene asked. "Do you think we need to request more DEA teams?"

"I'd say yes," Jim answered. "If all of the parlors are like the first one, it's going to take several more."

"What's your guess?" Gene asked.

"Well, there were six strip malls where the parlors were located, and we've asked for help on only one," Jim answered. "So, I'd say five more. And that doesn't include anything we find when we return to the farms."

"Guess I better let them know," Gene said, looking at the screen and locations of everything. "I just hope they have enough men here in Indy that we aren't left guarding the property until they can be brought in."

"I hope so, too," Jim replied. "And not to toss gas on the flame, but have you considered notifying ICE that we may need some support with the TDA folks in the two remaining apartments?

"And while you're thinking about it, we may need some local support to control the crowds when Mike has his men hitting the next apartment," Jim suggested. "Unless those TDA guys come out peacefully, which I doubt, you can expect a gun battle, and that will certainly draw attention.

"Looking back at the motel scenario, I'd expect at least as many people," Jim finished.

"You're right," Gene said, shaking his head. "And we had hoped to conduct this with as little publicity as possible."

"What's that line? Oh yes, 'silly rabbit, Trix are for kids'," Jim said as Gene took out his phone.

"I'm getting too damn old for this shit," Gene said dialing. "Maybe I'll give the big man back at Black Water a call and give him my resignation."

Chapter 50

Jim returned to Amanda's desk and asked, "Any luck on locating the car?"

"Just about," she answered. "Give me another minute."

"As soon as you locate her, pass the location to Minnie so she can track it," Jim told her.

"This would have been so easy if the phone that lady used had been in our system," Amanda said, watching the moving dot stabilizing in the area of Meridian and 86th Street. "Triangulation is the best I can do for now until Debbie plants a locator in her phone. But it's definitely headed north."

"We depend so much on technology these days," Jim observed. "What was state of the art a couple of years ago is akin to starting a fire with a flint stone today."

"You do know that most cigarette lighters still use flint, don't you?" Amanda asked. "Hang on, I see Debbie has got her phone transmitting. The car is about a mile south of 465. I'll highlight it so Minnie can get it quicker."

"Thanks," Jim told her as he turned to head for Minnie's desk.

"I've got it," Minnie was saying as Jim stopped beside her desk. "It looks like she may be following the same route as the first one."

"How long until she reaches Ming's house?" Jim asked, seeing the red dot crawling up the screen.

"Maybe ten minutes," Minnie answered. "Traffic should die out a little once she crosses 465. At least that's the way it seemed when you and Fabio were chasing the first car."

Looking at the screen, Jim made a quick call to Fabio and asked, "Anything happening there?"

"No, but Tammie called and said the buses should be here in a little over ten minutes," Fabio said. "And I'd guess the ambulances will get here a little before that since they're probably running lights and a siren."

"Shit," Jim said. "If that lady comes into the neighborhood and sees the action there at Ming's, she'll haul ass out of there.

"Minnie, get Amanda to tie my phone to yours and give me directions to Ming's," Jim said. "Keep me informed where the target car is and let me know if it doesn't continue toward Ming's."

"I'm on my way," Jim told Fabio as he sprinted for the door. "Maybe I can get close if they spot the activity at Ming's, and maybe I can block her way out of the neighborhood."

Exiting the hotel, Jim jumped into the Suburban he had left parked just outside the doors and started the car.

"Okay, Minnie," he said as he crossed Washington Street with horns blaring as he caused several west-bound cars to slam on their brakes to avoid hitting him. "I'm going to go left on Maryland and then north on Pennsylvania, is that correct?"

"Yes," she confirmed. "And if you remember, you should stay on Pennsylvania until you cross beneath I-65. Then make a left on 12th Street and then a quick right on Meridian."

"Got it," Jim said as he skidded around the corner to join Maryland, narrowly missing a car in the left lane. "Target still heading north on Meridian?"

"Yes, she just crossed under 465," Minnie told him. "She's about ten miles ahead of you."

"How far is she from Ming's house?" Jim asked, weaving back and forth across the lanes with his red and blue lights flashing and the siren screaming loudly.

"Maybe six miles," Minnie answered.

"Maybe I can get to her before she takes the exit at 146th street," Jim said, swearing under his breath at a car that pulled into his lane. "Do me a favor and call Fabio. I'd like to know if the ambulances are there.

"I'd call myself, but I don't dare with all of the drivers out here not paying a damn bit of attention to an obviously emergency vehicle," Jim said slamming on his brakes to avoid rearending a small car.

"Do you want me to ask Amanda to couple your phone with Fabio?" Minnie asked. "Then he can keep track of where you are and let you know when the ambulances get there."

"Please," Jim told her. "That would help. And when you get a chance, please tell General Barker where I am and what's happening."

"He's standing right here," she said, looking up at Gene. "I've got the speaker on my phone on so he can hear you."

"What the hell are you doing, Jim?" Gene said. "I thought you said the lady was heading for Ming's house and Fabio would handle it."

"Yes, sir," Jim replied. "That was until I figured out that she'd be getting there after the buses and ambulances. I didn't want her to disappear after going to all the trouble of finding her."

"What do you plan to do if you catch her?" Gene asked.

"I'm hoping I can trap her in the neighborhood once I get there," Jim answered, swerving around an eighteen-wheeler.

"And if you can't?" Gene asked.

"As long as Minnie can see both of us, she'll have to stop somewhere," Jim answered. "I don't think she'll have too many options at that point. Maybe she'll just go somewhere and park until she decides on her next move."

"Okay," Gene finally said. "But let's not try another of those moves that got you on the news last time you were in a car chase."

Chapter 51

As Jim was trying to get close enough to the Chinese lady, Mike was having problems of his own. Unable to break into the conversation between Minnie and Jim, he decided to call Butch and see if he could give him some fresh ideas.

"Hey, Butch, Mike here," he said as Butch answered.

"How's it going?" Butch responded. "Have you got those TDA folks from Venezuela under control?"

"The only thing I have under control is my frustration," Mike answered. "And I'm not sure if I'm controlling it or it is controlling me."

"What seems to be the problem?" Butch asked. "Is there something I can help you with?"

"Not unless you have a couple of tanks of Sarin gas," Mike replied. "I'm not sure if you saw the crowd at the motel where we made our first move on TDA."

"Nope, I didn't see or hear anything about," Butch answered.

"Let's just say our every move was potentially a political football," Mike told him. "We had the motel surrounded, just about to knock down the doors, and one of my guys guarding the rear was shot at by a couple of the targets.

"Then all hell broke loose," Mike explained. "But that's not what I called to talk to you about."

"I'm listening," Butch said as he watched his screen showing the teams going through the parlors.

"The thing I think upset Jim the most was the amount of press, and a bunch of people with nothing else to do but video everything," Mike explained.

"Now, I'm about to try and take out a bunch of TDAs in an apartment complex," Mike continued. "I'll bet these guys saw the motel shit on TV. The first apartment building went pretty smoothly, but I have two more, and at least one of them will hear about our raid on the motel and the first apartment.

"The big issue to me, given Jim's direction of a low profile, is that there is no way I can break down a shit load of doors, behind which are no telling how many weapons," Mike said. "And, given the number of people at the motel, and how fast the crowd gathered, I foresee a crowd scene like at a Rod Stewart concert."

"Any back door access?" Butch asked.

"Not really," Mike answered. "And even if there was one, I'm sure the selfie addicted rabble would find it. The biggest issue, along with the publicity, is getting the buses safely out of there.

"Let's not even talk about the bloodbath that will probably ensue," Mike continued. "Not just on their side, but possibly ours. It seems to me to be a disaster in the making. Hell, I can just envision a Rambo movie with millions of bullets flying everywhere."

"Doesn't sound good," Butch agreed. "Okay, I suggest you call Jim and get his advice."

"I've been trying, but he's out chasing some Chinese lady, I'm guessing one of your Triad people, and hasn't had time to get back to me," Mike explained. "I figured since he's helping you with your problem, you owe me at least some advice instead of calling him."

"Did you call and ask the General?" Butch suggested.

"No," Mike answered. "I'm sure he's got enough on his hands with the motel and the missing Triad folks. Not to mention the man who's supposed to be running this is out in the field instead of standing by to give us some direction."

"I've been with this organization exactly as long as you," Butch told him. "I only know that Jim must have been left with no other options, or he'd be back at the hotel.

"Now, I do have one suggestion, and I've already mentioned it," Butch started to say before being interrupted by Mike.

"No! I'm not running to the General," Mike repeated.

"Hold on, Marine," Butch said. "I'm as aware of the chain of command, military protocol, and crap as you are. But I also know that any Four-Star or Bird Colonel was once in your shoes … a tough decision and no simple way to accomplish the mission without violating some guidelines.

"I don't know exactly what the contract with the government entails," Butch continued. "But it seems as if the bottom line of your dilemma is deciding whether to remove the TDA or deal with a little bad publicity, I'd err toward removing the TDA.

"Hell, who can even remember the headlines from a week ago?" Butch said. "So a few political hacks get some egg on their faces. These guys are the masters at deflection. Teflon. Nothing sticks to them."

"So, your advice is to follow my original plan and let someone else take the shit storm," Mike asked.

"I've already made a suggestion, twice," Butch responded. "Call the General. Tell him exactly what you've told me. Lay out the options and then let him know you're in favor of pressing ahead regardless of the press or public opinion.

"There's one other thing you might ask when you talk to him," Butch said. "Ask if there are any local forces you can put in place before storming the beaches."

"I guess you're right," Mike said. "I'm here to corral the TDA folks because of the raping, burglary, murder, or whatever other despicable shit they've done, and make sure we get them out of the country."

Pausing for a moment, Mike said, "You know, if it came down to having my picture on the cover of Rolling Stone Magazine chopping off a few heads to protect my daughter from these animals, or standing in the shadows, I'd have a full-frontal face shot and a majestic profile of my beautiful face to accompany it.

"I'll make the call," Mike finally agreed. "And I'll present it the way I think it should go. If he objects, he can fire my fat ass and get me a ticket back to my little piece of heaven in Texas. At least on my ranch, I don't have to ask permission to blow some scumbag to hell and let the hogs eat what's left."

Chapter 52

"How far away is she?" Jim asked, crossing beneath I-65.

"About six miles," Minnie answered. "And she's about six miles from the exit at 146th Street. You're gaining on her."

"Yeah, I've gained four miles on her when she's driven four miles," Jim said, swerving around a carload of kids making faces out of the rear seat passenger window at him. "Using that ratio, I'll gain six miles while she drives six miles.

"In other words, I should catch her as she's taking the exit on 146th, which isn't too bad," Jim concluded. "But, if I'm as much as two minutes behind her, she'll be on Oak Road or maybe Maple Ridge. She'll see all of the flashing lights at Ming's house.

"She could then keep going straight and disappear in any direction," Jim explained. "I've got to get in front of her before that happens. Do you have any suggestions?"

"If you haven't caught up with her by 146th, I recommend you keep going north on Meridian and take the exit for 161st," Minnie suggested. "That way you'll be able to meet her head on, on Oak Road, if she doesn't turn into the subdivision.

"If she does turn in, she'll be on Maple Ridge and that's pretty much a dead-end road," Minnie explained. "Once in, the only way out is back to Oak Road.

"So, you'll have two shots at catching her if you continue past the 146th Street exit for less than half a mile, you'll exit right on 161st and Oak Road is less than half a mile on your right," Minnie continued. "So, you go to 161st, come south on Oak Road, and either meet her on that road, or come behind her if she enters the subdivision on Maple."

"That sounds like the way to go," Jim agreed as he continued to weave in and out of the traffic. "Plus, if she's on Maple before she sees the flashing lights, she may pull into the cul-de-sac, and I can block her in there.

"Plus, I'll have Fabio at the end of the road, and he can help," Jim added. "Do me a favor, Minnie, call Fabio and tell him to have all the emergency equipment turn off their lights when they get there.

"Better yet, if you can figure out what frequencies they are using, have them turn off all of the sirens and lights when they get on Oak Road," Jim continued. "Let's draw the fly to the spider's web."

"Jim," Gene suddenly said. "I just heard your plan, and I think I can help. I'll call the Chief of Emergency Services, and he can authorize running silently. The crews can't unilaterally turn off the equipment because of the liability the city would be exposed to. I'll let you know when I get it taken care of."

"Minnie, I still need you to call Fabio and explain the plan," Jim said. "I'd like to have him outside with his weapons just as a show of force."

"I'll take care of it," Minnie said. "And the car just took the 146th Street exit."

"Thanks," Jim said as he saw the sign for the 146th Street exit one mile ahead. "Did you ever play Pac-Man, Minnie?"

"No, but I have a nephew who loves it," she answered.

"Okay, you understand the premise of the yellow head catching the little, short-legged characters that look like small octopuses with half their legs cut off, and chomping them," Jim told her. "Let's pretend I'm the yellow head and you're directing me to chomp on the octopus car.

"Don't let me miss a turn or make a wrong turn," he continued. "And, since you can see her and see her options, try to stay one turn ahead of her by telling me which way to go.

"All of this may be moot if she doesn't see what's happening at Ming's house, but even if she gets close and then sees it, there are several other cul-de-sacs where she could run," he explained. "I want you to put me in a position to block her leaving that subdivision under any circumstances."

"Do I get one of those joy sticks to move you around?" Minnie asked, smiling to herself.

"Not today," Jim answered, passing the 146th Street exit, knowing his options were now down to one. "But I'll buy you a drink if this works out."

"I'd still rather play with the joystick," Minnie remarked. "By the way, Fabio said the ambulances just arrived, and they shut off their lights and sirens. But the buses aren't there yet.

"Okay, she's turning right on 151st Street," Minnie then said. "In half a mile, she'll be at Oak Road, and you should be at the exit for 161st."

"Looks like we may just trap her," Jim said, moving to the far-right lane for the exit.

Less than a minute later, Minnie said, "She's on Oak Road heading north. You're about a hundred feet from the 161st exit, and then half a mile to Oak Road on your right.

"Now, she's turning right on Maple Ridge," she told him. "She's in the subdivision, and the only way out is the road she's on.

"Oak Road is a hundred feet on your right," Minnie said. "Once you turn, you're about a mile from Maple Ridge. If you hurry, I don't think she'll even have a chance to turn around and make it back to Oak Road before you're there to block her."

"Starting to look like I owe you a drink," Jim said, accelerating on 161st. "Keep a close eye on her now. If she doesn't make the turn into Ming's cul-de-sac, let me know which one she's in if I turn on Maple Ridge before she can get back out."

"Not a problem," Minnie said. "She's slowed almost to a crawl. I guess she's looking for the street and house number. She's got about five hundred feet before she comes to the cul-de-sac on her right.

"There is one connecting street before reaching the end, but I doubt if she'd see it if she's looking down at the end of the cul-de-sac where Ming's house is," Minnie explained.

"You're half a mile from Maple Ridge," Minnie said, watching the red dot move slower and slower. "Once you turn left on Maple Ridge, you're about a thousand feet from Oak Bluff Drive on your right.

"She's just past the connecting street, so there's nowhere else to go except turn around and come toward you," Minnie told him.

"Okay, you're on Maple Ridge," she said excitedly. "Slow down, remember, you're only a thousand feet from a right turn on Oak Bluff.

As Jim turned right on Oak Bluff, Minnie said, "She's a hundred feet in front of you sitting still. What's your plan now?"

"I want you to tell Fabio to come down the street with his Bullpup showing, and I'm going to stop just behind her so she can't backup," Jim said, seeing the car parked beside the curb.

Pulling up behind the car, Jim inched forward until his bumper was almost touching hers.

Stepping from the Suburban, he saw Fabio coming toward them with the shotgun over his left shoulder. Jim walked to the driver's side front window and tapped on it with his Glock G18.

"Ma'am," he said as she rolled the window down, "Do you speak English?"

Hearing she did, he told her, "I want you and these girls to follow that man to the house at the end of the street. Do you understand?"

Again, hearing that she did, Jim reached down and opened her door just as Fabio stopped beside him.

"Please escort this lady to the house and put her with those kids in there," Jim said. "I'll bring these girls in to join them. Have the kids in there had anything to eat?"

"Not since we got here," Fabio said, helping the lady out of the car.

"I'm going to take a quick drive and get them some food," Jim said, letting the kids out of the backseat. "Take all of them in and put them together. The lady speaks English, so have her tell the kids I'm coming back with something to eat."

As Fabio led the lady and kids toward the house, Jim started back to his car and asked, "Minnie, does your map show me somewhere close to get some food for these kids?"

"Just a second," she answered. "Okay, there's a place called Kizuki Ramen about a quarter of a mile south on Oak Road. Will that do?"

"At this point, I bet those kids would eat dirt, hoping there was a worm in it," Jim said, turning around to leave. "Now, give me directions to this noodle place and see if you can find out what the hell is taking the buses so long to get here."

Chapter 53

Jim dropped off the food for the kids and was just getting back in the Suburban when two buses pulled up to the curb outside Ming's house. Getting into his car, he took out his phone and called Mike.

When he answered, he said, "Sorry, I've been sort of out of touch for a few minutes. What's happening at the TDA apartments?"

"We've taken the second apartment," he said. "It wasn't pretty, and I'd bet there's two hours of videos for the news to show.

"The good news is that we only had to put one man down, wounded a couple more, and didn't get any of my men hit," Mike told him. "I'd say that's pretty successful."

"Sounds good," Jim replied, heading for Oak Road. "Guess I'll take some heat for the public display of the ruthless ICE, though."

"I covered that," Mike assured him. "I talked to the General and explained the situation. But, before I called him, I discussed the situation with Butch, and he agreed that our job was to get the TDA.

"Anyway, the General said he'd cover my ass regardless," Mike told him. "Guess he's one of the good guys. My experience has been that they'll toss you to the wolves if the operation doesn't look favorably for them.

"I have to admit, it's nice to work for a guy who truly has your back, even if it looks bad for him," Mike finished.

"Yeah, he's that kind of guy," Jim replied, heading south on Oak Road. "What's your plan on the last apartment?"

"Pretty much the same," Mike answered. "The guys will be busting in their doors in less than a minute now. And you wouldn't believe the crowd of cops, reporters, and looky-loos.

"You'd think the President was about to make an appearance," he continued. "But I'm hoping that we'll be done in the next thirty minutes or so."

"Great," Jim told him. "If you want to go lend a hand as they escort the last TDA from the building in cuffs, you might get to be a guest of one of the news shows. Maybe one of the reporters on site will want to interview you."

"No, thanks," Mike told him. "I'd just as soon remain the anonymous man behind the scenes. But if you want to come and say a few words, I can give you the address. You can probably get here just in time to march the last man out to the buses."

"I think I've had my share of public buffoonery," Jim replied, turning onto 151st Street. Anyway, I apologize for leaving you to fend for yourself."

"Not a problem," Mike told him. "By the way, how'd the chase of the Chinese lady turn out?"

"No reporters, no crowds, no bystanders, just a couple of ambulances and buses to take out the trash," Jim answered. "Again, I'm sorry I wasn't there to help you. But it looks like your guys did a good job. Let me know how it goes with the last apartment."

"For sure," Mike told him. "But you really owe me for this."

"I have no doubt that you'll be presenting the bill sooner, rather than later," Jim said, turning on Meridian.

Merging with the southbound traffic, he called Butch and asked, "Butch, how's it going at the rest of the parlors and the farms?"

"Pretty much the same as the first one," Butch answered. "We've pretty much turned it over to the DEA.

"I've heard from most of the farms, and we haven't found any drugs or shit. Looks like they stuck pretty much with just the weed," he finished.

"I can't say that doesn't surprise me," Jim replied. "Maybe they were a little worried that if they got raided by the DEA, it wouldn't be as bad as if they found drugs there."

"Sounds reasonable," Butch agreed. "Plus, there probably wasn't the sales opportunity there like in the parlors. They were strictly a production facility and left the distribution to the parlors or street vendors."

"Anyway, I'm headed back to the hotel and see what we need to do to wrap this little shit show up," Jim said. "I'm mentally preparing myself for a major ass kicking. I know the General is going to have a lot of explaining to do to his boss, and he'll take most of the heat. But this is not the way the company wants operations to be handled."

"Good luck, sir," Butch replied. "You know, I've sort of enjoyed this. Maybe not as much as Mike, but it feels good to be taking those worthless assholes who exploit kids off the streets and out of the country. Especially those involved with the sex stuff with those little girls. Really good about what we're doing."

"I agree," Jim said. "I don't know how it had to get this bad before something was done. Maybe that's why I'm not a politician … I can't keep my mouth shut when shit like this is going on.

"Anyway, I guess we'll get the final word this afternoon when it's over," Jim told him. "I'll be back at the hotel in a few minutes, and if you need anything, I hope I'm there instead of chasing down someone who slipped through our fingers."

"About that," Butch responded. "That's my fault. My people screwed up."

"No, it's not anyone's fault," Jim told him. "Nobody can see everything when you're dealing with shit like this. Hell, if it were easy, they could send in the Coast Guard.

"Don't even think you didn't do your job as well as anyone could have," Jim told him. "I'm glad I recommended you for the job."

"You recommended me?" Butch asked. "Then you really are an asshole. Doesn't the company have any operations in Bermuda, or maybe the Florida Keys, to make sure the bikinis are small enough? How about next time, recommending me for that?"

"I'll keep that in mind," Jim answered. "See you at the hotel when we're done."

"Before you go," Butch said. "How's Jewell?"

"Pretty good, considering," Jim answered. "I'm going to stop in and check on her before I get back to the hotel. I'm just a mile or so from the hospital now."

"Good," Butch replied. "Please tell her I'm sorry she got surprised with the knife thing and I don't blame her a bit."

"I'll do that," Jim said, seeing the hospital just ahead. "Let me know if there's anything you need to wrap up the farmer and parlor thing."

Chapter 54

As Jim was pulling into the hospital parking lot, his phone rang. "Hello," he answered.

"Hey, Jim," Fabio said. "I just got another call on Ming's phone. What do you want me to do?"

"What did the caller want?" Jim asked.

"They didn't say anything," Fabio told him. "Of course I didn't answer, hoping they would call back."

"Give Amanda a call and have her connect you with Debbie back in Quantico," Jim told him, getting out of his car. "Then she'll tell what she needs to find the owner and put a tracer bug in the phone."

"Got it," Fabio said. "Do you have any idea when someone can come get me?"

"Are the ambulances gone and buses loaded?" Jim asked.

"The ambulances have left," he answered. "They are taking the body to some hospital around here to make the death pronouncement and say we can go there to claim the body."

"This *can't pronounce them dead shit* is getting to be a pain," Jim said walking into the hospital. "When half of a man's guts are sticking to the wall behind him and he's as cold as your ex-wife's heart, the son of a bitch is dead.

"I just don't understand why trained medical people can't make an obvious call," Jim continued as he headed for the room where he last saw Jewell. "Anyway, call Amanda and talk to Debbie. Then let me know when the buses are gone. If I'm still here at the hospital with Jewell, I'll come get you."

"Yes, sir," Fabio said. "Do you want me to call and tell you what Debbie wants?"

"Not necessary," Jim answered, approaching Jewell's room. "I'll either hear from her, or you can tell me when I come to get you … if I come get you."

"I thought you just said you'd come from the hospital back to get me," Fabio replied.

"Pay attention," Jim told him as he almost ran into the doctor coming out of Jewell's room. "I said if the buses were gone before I leave the hospital. Now, I've got to go."

"Doctor, may I have a moment?" Jim said, hanging up his phone.

"How can I help you?" the doctor answered, looking at his watch.

"The patient in that room," Jim said, nodding to Jewell's room, "works with me and is also a close friend. Can you tell me how she's doing?"

"I'm guessing we'll be releasing her tomorrow morning," he answered. "She had a nasty cut on the left side of her head, and we got it stitched. Her X-rays and MRI don't show any permanent damage, no brain swelling, or fractures.

"I've recommended keeping her for another twelve hours or so just to monitor her in the event there's some bleeding within the brain that hasn't manifested observable symptoms yet. But we'll know in the morning.

"Is there anything else?" the doctor asked. "I've got a pretty full schedule right now, but I can spare another minute or so."

"No, sir," Jim answered. "I appreciate you taking the time to talk to me. And I appreciate everything you've done for Jewell."

Jim stepped to the door of her room and saw her sitting up, reading a newspaper. "Looking for apartments for rent in the Indianapolis area?" he asked. "I've heard that there are several vacancies right now."

Laying her paper down on the bed, she said, "No, I think I'll just go back to Chicago. From what I saw on the news, when they let me watch that little TV up there in the corner, a lot of those apartments might possibly be biohazards."

"Nothing a little Lysol and a good scrub brush can't handle," Jim told her, stepping beside her bed. "How are you feeling?"

"Minor headache," she answered. "But I'm guessing that's due to the wonderful chemicals they have running through my veins. Give me another ten to twelve hours, and I'm sure I can work up a hell of a headache."

"Not to mention the itching that comes with having stitches," Jim replied, smiling at her. "And it looks like you've had one of those modern haircuts where they shave a design in the side of your head resembling a lightning bolt. Very stylish.

"Anyway, I wanted to stop by and check on you before I get back to the salt mines," Jim said. "And I talked to your doctor, and he thinks you can go home tomorrow morning. That is if you don't get the big head."

"I don't see why I can't leave right now," she complained. "I feel fine, I don't have blood pouring out of my head, and I want to go back to my hotel room and take a hot bath. Can't you sneak me out?"

"I'm in enough shit right now," Jim answered looking at the headlines of the newspaper laying on the bed.

Picking it up, he pointed to the photo of the overturned car with himself face to face with the motorist who wouldn't leave, saying, "I know I can't be identified from this picture, but I know it's me. And Gene knows it's me. Therefore, Black Water knows it's me. My low profile has risen to stratospheric levels. Not what we're supposed to do."

"Speaking of that, what happened to the lady and the kids?" she asked.

"The lady died in the accident," Jim answered. "The kids got knocked around a bit, but the EMTs thought it wasn't life-threatening or anything.

"I believe they might be here in this hospital," he continued. "You might be able to ask one of the nurses.

"Oh, forget what I said about the lady dying," Jim said as an afterthought. "I'm not a medical doctor, so I'm not qualified to make the call."

As his phone rang, he looked at the caller ID and told Jewell, "You just rest, and maybe I'll get to see you in the morning before I head back to Texas.

"Now, I'm still trying to wrap up this shit show and maybe the company won't spank me too hard," Jim said turning to leave. "I'd tell you to take two aspirins and call me in the morning, but as I said, I'm not a doctor."

"And your bedside manner could definitely use a little improvement," Jewell said as Jim headed for the door. "Maybe you can come back later tonight, and I'll try to teach you."

Jim simply turned to look at her and shook his head as he left the room.

Chapter 55

"Yeah, Fabio," Jim said as his call was answered. "What did you find out?"

"I talked to Debbie, and she asked me to call the number back that called Ming's phone while she was monitoring it," Fabio answered.

"How'd that go?" Jim asked as he got in his car.

"Good, I guess," Fabio answered. "When the phone was answered, she made some kind of static on the phone, then hung up.

"A minute later, the phone rang again," he continued. "Then Minnie started talking in Mandarin. They spoke for a few minutes, and Minnie hung up."

"Did anyone tell you what was said?" Jim asked as he started the car.

"No," Fabio answered. "I guess they're waiting for you or Butch before they let anyone know. By the way, all of the buses are gone. Are you coming to get me?"

"I'm on my way now," Jim answered as he headed back to Meridian. "Should be there in ten or fifteen minutes. But

I'd better go now and see what's happening with the mystery caller."

Heading north on Meridian, he called Minnie and said, "I understand we have a new player in the never-surprising Triad saga."

"I'm afraid so," Minnie told him. "I was just about to call you. Debbie has pinpointed the location, and it's at an apartment on the north side of the city, fairly close to you.

"I talked to Butch, and most of his people are headed back out to the farms as you directed," she continued. "So, I figured I'd see if you wanted to go take a look at what's going on at that apartment. I knew you were still at the hospital and figured you could get there faster than anyone else."

"Okay, I'll head over there after I pick up Fabio," Jim said, turning on the flashing lights and siren. "I think it might help if we have Ming's phone.

"Now, can you check with Debbie to see if she can combine your phone with Ming's so you can call from it and speak directly to whoever answers?" Jim asked, weaving through the traffic.

"I'll give her a call," Minnie replied. "Do you need me to guide you to Ming's house again?"

"Hell no, I think I can find it blindfolded now," Jim answered, pulling up on the bumper of a car that didn't seem to notice the flashing red and blue lights behind him.

"Do me a favor and let Gene know where I'm going," Jim said as the slow-moving car finally pulled over to let him pass.

"I'm right here, Jim," Gene said. "It seems that I've inherited your job since you decided to chase that first car."

Pausing to see if Jim would respond, Gene then said, "At least this last one didn't make the news. I'm damned grateful for that. And I can see the urgency in getting someone to the apartment where the call came from.

"I'm starting to wonder how many other people are hiding out somewhere that we don't know about," Gene concluded.

"Maybe I can get some answers from whoever made the call," Jim said, taking the exit for 146th Street.

"I hope so," Gene responded. "This whack-a-mole game could go on forever if we keep having new players popping up every time we seem about to wind things down."

"I agree," Jim said, turning east on 151st Street. "I'll have Fabio and Ming's phone in a couple of minutes, and we'll head to the apartments.

"And I'm going to need directions, if Minnie is still listening," Jim said, hitting the roundabout just prior to Oak Road.

"I've already got it mapped out," Minnie said. "I'll start giving you directions as soon as I see you leaving Ming's house."

"That works for me," Jim replied, turning north on Oak Road. "How far away is it?"

"Looks like a little over ten miles," Minnie said. "Probably just a couple of miles east of Meridian on 465."

"I'll grab Fabio and hurry over there," Jim said, pulling onto Maple Ridge. "Please give him a quick call and tell him I'm less than a minute away."

"Doing it now," Minnie said.

"Do you think we need to send someone over to Ming's house since you're taking Fabio?" Gene asked.

"I don't think so," Jim said, turning on Oak Bluffs Drive. "I'll ask Fabio, but he was supposed to do a thorough inspection of the house while we waited for the ambulances and buses. But I'll verify that as soon as he gets in the car.

"You might want to notify the DEA," Jim added. "Since Ming was involved with the parlors where the drugs were found, maybe they can use that as reasonable cause to search the house. Toss in a Rico charge, and you'd have a mighty big case against the Triad."

As Jim came to a stop in front of Ming's house, Fabio stepped from the driveway and pulled the car door open, saying, "I'm glad you finally got here."

Sliding into the passenger seat, he asked, "Aren't I supposed to be the driver?"

"Yeah, and I'm supposed to be standing behind Minnie drinking a Dr. Pepper and eating Ritz Cheese Crackers," Jim said, accelerating back the way he had just come.

"Not to ask what you've been doing while I've been doing your job, but did you manage to search the entire house?" Jim asked, heading west on 151st Street.

"Yes, sir," Fabio assured him. "I went through every room and checked for hidden areas. Nothing except what we were already dealing with.

"And, not to criticize the way things are going, but I'm starting to get a little hungry," he announced.

"Didn't you eat any of the stuff I brought for the kids?" Jim asked with a quick glance.

"Hell, I was afraid to even reach for a noodle," Fabio answered. "There had to be enough food for fifty people in those sacks, but those kids went through it as if they hadn't eaten in a year.

"Like I said, I was afraid to reach for anything," he continued. "I was afraid one of them would stick my fingers in the wasabi sauce and chow down. Poor kids."

"Well, maybe there will be something left of the lunch buffet when we get back to the hotel," Jim said, merging with the traffic on Meridian.

Chapter 56

No sooner had Jim got to Meridian, Minnie said, "Keep heading south. Stay in the right lane for the exit for 465.

"Head east on 465 when you get to it," Minnie told him. "You'll go about two and a half to three miles and take the Keystone Parkway exit. Stay on Keystone going north for about a quarter mile and look for the East 96th Street exit. Go east on 96th Street and take a left at Allisonville Road."

"Slow it down," Jim said. "I can't remember all of that and dodge traffic. Let's go back to you telling me which lane to be in and the next road. That's about the extent of my memory at this point. And watch my little green dot to make sure I'm slowing down before I pass the turn.

"The way this city is laid out, it could be five miles before I can return to the road I wanted. Just let me know when I get close to whatever road I need to take," Jim told her as he got on 465.

"But, while you're waiting to tell me where to go, can you see if Debbie knows who owns the apartment where the call originated?"

"I'll ask her," Minnie answered. "You're still two miles before the next turn."

A couple of minutes later, Minnie said, "Stay in the right lane. The exit for Keystone Street will be less than half a mile. Stay in the right lane for less than a quarter mile and go left on Keystone."

"Got it," Jim said, looking for the exit sign.

"And I got the information you asked for," Minnie said. "The apartment complex is owned by Liming International Apartments, Inc."

"Are you saying Lemming, as in the rodent?" Jim asked as he took the exit to Keystone.

"I don't know," Minnie said. "Hold on. And your turn onto 96th is the next road on your right."

Making the right turn on 96th, Jim asked, "What's the next turn?"

"North on Allisonville Road, about four miles in front of you," she answered. "And Debbie says it's spelled L-i-m-i-n-g. Maybe like in lime-ing, you know, the fruit, lime?"

"No, I don't think so," Jim told her. "I'd bet it's more like Li Ming, or Ming Li, the guy whose house we just left."

"Oh, I guess you could be right," she said. "After you turn north on Allisonville, about two more miles in front of you, the apartments are on the left side of Allisonville."

"Got it," Jim told her. "Are you still connected to Ming's phone? I have a feeling we're going to need a Mandarin speaker when we get to the apartment."

"I believe so, let me check with Amanda," she answered. "If not, I'm sure she can re-establish the connection."

A moment later, Minnie said, "Yes, I'm still connected. And you're almost to Allisonville. Go left on Allisonville.

Slow down on Allisonville because it's only half a mile to the next turn."

Hearing how close he was, Jim turned off the lights and the siren to avoid alerting whoever had called Ming's phone.

Turning left on Allisonville, he said, "I'm going to ask you to get on the phone, the Ming connection phone, and call that number when I tell you to. I want you to be on the phone with them when I knock on their door.

"Your job then becomes convincing them that they need to let me in or talk to me," Jim told her.

"Okay, maybe I think I can make them think you're with Mr. Ming," she told him. "And Deerbrook Drive is the next street on your left. The apartment is about two hundred feet down on the left. Corner of Deerbrook Drive and Falling Brook."

Turning onto Deerbrook, Jim looked around to see what sort of neighborhood they were in. It appeared to be a mix of lower-rent apartments and various businesses.

Deerbrook seemed almost void of any cars along the street as far as he could see. Spotting where he thought the apartment was, he asked, "Is that it on my left right now?"

"Yeah, that's it," Minnie answered. "When do you want me to make the call?"

"I'm going to drive past and then turn left to see if there are any cars parked around the apartment," Jim answered as they cruised by slowly looking for any place a car from the apartments could be parked.

Turning left on Falling Brook, he looked at the back of the apartments and asked Fabio, "Do you see any parking spaces for these apartments?"

"No, and that's sort of strange," Fabio answered, checking both sides of the street.

"I know," Jim agreed, going to a driveway about a hundred feet past the apartments. "I don't know how anyone living there could go anywhere. Maybe a taxi, but that doesn't seem likely in this neighborhood.

"And I didn't see any bus stops," he continued, turning around to go back to Deerbrook. "This is a strange situation. When we get back, I'll take the door, and you stay to my left with your Bullpup ready to shoot through the crack when they open the door."

"Why don't you stand to the side and let me knock on the door?" Fabio asked.

"Because I may need a driver to get me to the hospital," Jim answered with a quick glance at him. "And, if I'm not mistaken, you are my driver."

Pulling to the curb in front of the apartments, Jim told Minnie, "Call the number. When they answer, I'll be listening. Please convince them that I'm there to help them."

"I'll do my best," she said. "I'm calling now."

Chapter 57

Jim pulled off his jacket and vest as he got out of the car and laid them in the seat.

"What the hell are you doing?" Fabio asked joining him on the driver's side. "You don't plan on going in there without your vest and helmet, do you?"

"I don't want those kids to see Darth Vader standing there when they open the door," Jim explained. "They're probably already scared shitless."

"What if it's some man with an Uzi pointing at you when he opens the door?" Fabio asked as Jim started up to the building.

"If there was an adult in there, why has a little girl been the only one using the phone?" Jim asked as they got to the door.

Looking at the door and noticing it would open on the left side, he said, "Take the right side so you can see into the apartment. If there's a gorilla standing there when it opens, feel free to start pumping shells through the crack until there's two hundred pounds of raw hamburger on the floor."

Hearing Minnie talking to someone and hearing a girl's voice answering, he knocked on the door. Waiting and looking at Fabio's face, he whispered, "It's still a little girl. But be prepared just in case."

The door slowly opened, and Jim saw a little girl, probably no more than twelve, standing there with the phone to her ear. He knelt down and smiled at her.

"Tell her your name is Jim," he heard Minnie telling him. "I've told her you are there to take her home."

Jim looked into the girl's eyes and softly said, "I'm Jim. I'm going to help you."

The girl looked behind her and said something Jim didn't understand, but Minnie said, "Her name is Mei. M-e-i. Pronounced Ma, like your mother."

"Hello, Ma," may I come in?" Jim asked as Minnie interpreted.

The little girl stood quietly as Minnie continued to talk to her, but finally opened the door and bowed slightly.

Jim bowed in return and slowly rose. "Stay here," he told Fabio before stepping into the apartment.

Once inside, he saw several girls standing in the room looking at him. Bowing slightly to them, he asked Minnie to see how many kids were in the apartment.

When she finished the conversation, Minnie told him, "There are ten in this one. And ten more in each of the five apartments on that floor. There are also some boys in the apartments on the floor above their apartments, but she doesn't know how many."

"Ask her if there are any adults here," Jim said.

After a brief exchange, Minnie told him, "There were two this morning, but they left."

Trying to figure out who the men were, Jim asked Minnie if she had the pictures of the two men at Ming's. Hearing she could get them; he asked her to send the pictures to the little girl's phone.

A couple of minutes later, Jim heard Minnie talking to her, and she looked at the phone.

"Those are the men, she says," Minnie told Jim.

"Good," Jim said. "At least we know they won't be coming back. Now, please have Tammie get a couple of buses headed this way. That's about one hundred kids if the top floor has an equal number of boys as this floor has girls.

"And make sure there are Mandarin speakers on those buses," Jim directed, looking at all the little girls watching him.

"I'm going to stay here until the buses arrive," Jim said. "Let's not take a chance of losing the slight trust we've got. Just get the damn buses here as soon as possible.

"Ask Tammie if she might get the buses that have already dropped off their loads at the Feds," Jim suggested. "Find the closest ones or send forty cabs with speakers in them. I want to get these kids out of here before Ming's assistant, if there is one, figures out what's happening and comes to get them."

"What do you want me to do?" Fabio asked with his shotgun held out of sight behind his back.

"Go back to the car and take off all the ICE stuff, leave your guns, and then I'll let Minnie tell these kids you're with me," Jim told him.

"Once they understand, I'm going upstairs and see if I can get the boys to accept me," Jim finished.

"Did you hear that?" Jim asked Minnie. "Do you have any suggestions that may help?"

"Maybe," she answered. "Let me see if I can convince Mei to go with you. That may help."

"Sounds good," Jim said, looking at the filthy dresses the girls were wearing. "Is there anything else you can ask? Like, where did they come from? How did they get here? How long have they been here?

"Anything just to gain their trust," Jim said. "As long as they are talking, they won't be thinking about who the hell I am or what's going to be happening to them."

For the next few minutes, Minnie carried on a running conversation until she told Jim, "Mei says they came from different provinces in China. They were taken from their homes a couple of weeks ago and traveled down here from Canada.

"They got here two days ago and are waiting to go to their new homes," Minnie explained.

"And they all want to go home," she finished.

When Fabio returned, Jim asked Minnie if it was all right to leave him there and go up to see the boys. Hearing that Mei was ready and not scared, she said something to Mei and then told Jim to take her hand, and she would lead him.

Jim reached down and took her tiny hand in his and gave a slight bow. As she led him outside and up the stairs, he wondered how low a man would have to be to treat anyone like this. Especially exploiting the little girls for what amounted to slavery and underage sexual assault.

Chapter 58

About thirty minutes later, two yellow school buses pulled to the curb behind Jim's car. As the doors opened, a lady got off each bus and came up the walk toward the apartment.

Jim stood just outside the apartment to meet them and explained that there were boys on the top floor and girls on the bottom one. Probably about one hundred kids in all.

Mei had been reluctant to let go of Jim's hand after they had come down from seeing the young boys, and Jim seemed happy to let her hold on.

"This little lady is Mei," he said to the first lady from the bus. "She's from Shandong and misses her father, who is a fisherman.

"And no, I don't speak Mandarin. But Minnie, the lady who has been my interpreter, translated for me," Jim continued, looking down at Mei and smiling. "Mei has been telling me more about her family and what she misses than I ever thought possible from a stranger. Especially a kid.

"But I guess she feels the need to talk to someone," Jim finished. "I guess that someone is me right now.

"I think it may be better to take the boys in one bus and the girls in the other," Jim suggested. "I say this because disrupting their tentative bonding could just add to the stress of being stolen from their families. But I'll be glad to turn them over to you ladies."

Jim knelt down beside Mei and asked Minnie to translate as he tried to explain that he had to go so he could find other kids who also needed to be taken home.

After a prolonged hug from Mei, Jim stood and told the lady standing beside them, "This is my new friend, and I want you to take special care of her. She helped me convince the rest of them that we were here to help them."

As Jim turned to walk away, Mei grabbed his leg and wrapped her arms around it, crying something he didn't understand.

"She says she wants to go with you so she can tell the other kids that Buddha sent you to take all of them home," Minnie translated. "I think the kid likes you, Jim."

"Tell her that I have to travel far and fast and her legs can't keep up," Jim told the lady standing beside them. "But I will do my best to make sure all of the children get home soon."

Gently pushing her away, Jim said, "Mei, you have to go with this lady and all your friends who were brought here. Just remember that sometimes we have to let go to find happiness."

Signaling to Fabio it was time to go, Jim bowed to Mei and nodded to the lady with her saying, "I wish I could stay and help you get these kids to wherever they are going to be taken, but I'll leave it to you guys to try to let them know they will eventually be taken home."

"Let's get going," Jim said, turning away and heading for their car. "It's time for you to do your job, driving me, instead of standing around playing games with a bunch of kids."

"Don't try that tough guy shit with me," Fabio said as he opened the driver's door. "I just saw what's beneath that hard-ass attitude you portray. And I thought you Marines were the biggest, baddest creatures to ever roam the earth."

"A man never stands so tall as when he kneels to help a child," Jim replied. "And never confuse compassion with weakness. Now, you just listen to Minnie, and she will get you back to the lunch buffet."

"Tell me where to go, Minnie," Fabio said as he started down Deerbrook.

"Right turn on Allisonville Road," she told him.

"By the way, I didn't know you knew any profound Buddhist sayings, Jim," Minnie remarked.

"What are you talking about?" Jim said as he saw the intersection with 96th Street just ahead.

"That part about letting go to find happiness," Minnie answered. "That's one of the more well-known sayings. Both Buddha and Thich Nhat Hanh tell you that sometimes you have to let go to find true happiness.

"By the way, turn right on 96th," she told them. "Then 465 to Meridian. I hope you can find your way back to the hotel from there."

Jim looked at Fabio, shaking his head, and asked, "What? Is there something you don't understand?"

"Yeah," he said, turning onto 96th. "How do you know these Buddhist sayings?

"I don't know crap about Buddhist philosophy," Jim answered. "Hell, I was referring to the happiness a man feels after the final divorce decree."

"I heard that, Jim," Minnie said. "And true happiness only comes after the final alimony payment. That's according to my ex-husband, anyway.

"Now, Butch has been trying to contact you. If you don't need my assistance, I'll disconnect the hot mic thing and let you see what he needs."

"Thanks, Minnie," Jim said. "Not just for the directions, we certainly needed to get everywhere we've had to go. But more specifically, for helping with those kids back at the apartments. I don't think we could have taken care of it if you hadn't convinced Mei to trust me. Thanks."

Chapter 59

As Fabio made the turn south on Keystone, Jim called Butch to see how much progress had been made regarding the search of the farms.

"How's the corn crop this year?" Jim asked when Butch answered. "I've heard that the price of corn is really popping."

"That's terrible," Butch replied. "Save your jokes for those kids you're finding, if that's the best you can do.

"But, to answer the underlying question, it's going pretty well," he answered. "The guys finished the first farm, and there wasn't anything remarkable out there. They went through every room, crevice, closet, and looked behind every picture on the walls, and nothing.

"They did the same at one of the outbuildings," he continued. "A tractor, an old pickup truck, cases of motor oil, a few tools, but nothing you wouldn't expect to find. I'm waiting for the next team to let me know what they've found. But honestly, I think this is a waste of time."

"You could be right," Jim said. "But my biggest fear regarding the farms is missing one of the kids. I'm not too worried about the drug issue; hell, we can turn that over to the DEA. Since marijuana production is illegal in Indiana, they can go to those farms we've identified and search every square inch.

"We are here to clean up the Triad operation," he continued as Fabio merged onto 465 heading west. "I just came from a location where it looks like the Triad brings their fresh recruits, at least for this area. I don't know how long they are kept there until farmed out, but I'd hate to think we have a couple of twelve-year-old boys hiding in one of the farms.

"The kids I saw have only been in the country a couple of days, and don't speak a word of English," Jim told him. "Just think of how you'd feel if you were left alone, possibly locked in some room, and afraid of being caught by the Gweilo, who you've been told will make you a slave for the rest of your life."

"What's a Gweilo?" Butch asked.

"A white man," Jim answered. "White devil. Gweilo is an insulting name for people the Chinese hate. Much like the terms used to describe the blacks here in America. Or Hispanics. Hell, even the Irish were ridiculed some years ago.

"Probably goes back to when the Chinese, blacks, Native Americans, and Irish were forced to work on the transcontinental railroad for less pay than their white counterparts," Jim continued.

"What I'm trying to say is, these kids are probably so scared of us that they'd trust the men who brought them here instead," he explained. "Like the Stockholm Syndrome.

"If we don't find these kids, they have few options other than being on the streets and probably end up in some gang,"

Jim said. "Our primary goal was to shut down the Triad's operations here in this country, and we've pretty much done that here in Indianapolis.

"I think we have a responsibility for the collateral issues, those kids who may be abandoned," Jim continued. "Maybe I'm overstepping my directions from the company, but this has turned into more than a slash-and-burn operation for me. I don't think anyone can see what's happening to these kids and not feel some empathy.

"Would we feel differently if they were white children?" Jim asked. "I hope not. Kids are kids. Anyway, let's take a few extra minutes to look at those other farms."

"Not a problem," Butch said. "We've only got a couple of the businesses located with the massage parlors to look at, and we'll be done there. I can send those guys out to the last of the farms when they're done. That should get this finished within the next hour or so."

"Sounds okay to me," Jim said as Fabio joined Meridian heading south. "I was a little overly optimistic when I hoped to have this entire operation completed by noon. But I think the tough part is done and just the cleanup remains."

"Have you talked to Mike lately?" Butch asked. "I know he was about to hit the last apartment last time I talked to him."

"To be honest, no, I haven't," Jim answered. "Between your Chinese madams who got loose, finding who could be the number one man for the Triad here in Indianapolis, discovering a storage facility for new arrivals from China, I've sort of left the TDA issue to Mike.

"I'm heading back to the hotel and should be there in the next ten minutes or so," Jim told him. "I'll have a chance to catch up on his operations and watch as your folks wrap up all the areas we've been discussing.

"If Mike has managed to clean out the last apartment, I'll see if he can spare a few men to help you at the farms," Jim finished. "I'll give him a call right now and see how that's going."

As they crossed over the White River about six miles north of the hotel, Jim called Mike and waited for him to answer.

"Hello, Jim," Mike said a moment later. "Are you calling to tell me you've found another reason to stay away from the hotel or to tell me you've decided to join me for the press conference when we finish going through the last room here at this apartment building?"

"Wouldn't that be pretty much the same thing?" Jim asked. "A press conference at the apartment would certainly be a reason to avoid the hotel. I think I'll go to the hotel to avoid a press conference.

"Anyway, I've got to get back there and see if we have the slightest chance to conclude this before some group of human rights wackos brings a hundred well-meaning demonstrators to express their opinion regarding the civil rights of illegal immigrants," Jim finished.

"I think you're a little late for that," Mike told him. "I thought the fiasco at the motel was epic. That was a five-year-old's birthday party with the neighbor kids compared to what I have going on here now.

"I knew these last two apartments wouldn't go as smooth as the first one we hit after the motel," Mike continued. "But I don't know where all these people are coming from.

"There are news vans from every network, hundreds of protestors holding signs saying there are no illegals in stolen lands, whatever the hell that means," Mike said. "I'm starting to worry about getting my men out of here safely.

"Wouldn't that be the shits to survive a gun battle with these TDA animals and then be clubbed to death with a sign by some numb-nuts whose IQ is somewhere in the vicinity of a potato?" Mike asked.

"Not to redirect the conversation," Jim said. "What is your realistic estimate of when you'll have that apartment cleared?"

"Twenty minutes, tops," Mike answered. "I've actually been somewhat surprised at how quickly these last ones have been. I don't know if it's because of what they saw happening at the apartment building across the parking lot or on the news, but most of the apartments here have opened their doors when we knocked.

"There was only one where we had to knock the door down, and we haven't had to use a flashbang since the motel," Mike told him. "Whatever convinced these guys to lay down their weapons is great as far as I'm concerned. Even if it was the publicity, which I know bothered you.

"But there hasn't been a single shot taken at my guys since the first apartment," Mike finished. "I never thought we'd get this far without some casualties. If I'd known the coverage by the media would have this effect, I'd have conducted a press conference early this morning."

"I agree," Jim said as they pulled up to the hotel. "Doesn't matter what caused the change in their attitudes. The results are what matter. I'll see you here as soon as you wrap up that last apartment building and finish your press conference."

Chapter 60

"Where do you want me to leave the car, sir?" Fabio asked as Jim got out at the hotel entrance.

"Just park it somewhere close that doesn't block the vans or cars that are unloading hotel guests," Jim answered. "And make sure all of the weapons are checked, reloaded as necessary, and put out of sight."

"Got it," Fabio said. "I'll see you inside in a few minutes."

Walking into the conference room, Jim saw Gene standing behind Minnie and talking on his phone. Standing to the side and looking at the screens, he wasn't surprised to see the number of green icons and the few red ones. Other than the mass of red dots at the final apartment, there were no others at the parlors, farms, or other apartments.

"Welcome back," Gene said, hanging up. "How was your adventurous vacation?"

"I highly recommend it," Jim answered. "A chance for exciting high-speed car chases, a chance to meet some very unusual characters, though highly distasteful, attended the coming-out party for some delightful twelve-year-olds, and did I mention high-speed car chases?”

“Yes, you did,” Gene replied.

“Well, I thought I’d mention it twice because I had the pleasure twice,” Jim said, nodding. “How did you enjoy being the kingpin of an international drug and child smuggling takedown?”

“I think I would have enjoyed it more if I hadn’t been in the middle of it,” Gene told him. “I do believe that was exactly why I brought you here.”

“My apologies,” Jim responded. “I was under the impression I was here to manage the capture of TDA and Triad members.

“Using the only available resources at the time of discovering targets the company failed to identify, or that escaped before they were captured, I assigned the most capable member to make the capture.

“And given how quickly this operation was plummeting into the lowest level of the Porta Potty, I reassigned myself,” he finished.

“You never fail to amaze me, Jim,” Gene said. “I swear you could reverse any possible argument regarding what you were assigned to do versus what you actually did into a potential endorsement for a Presidential Citation.

“Now, that’s a very unique talent,” he continued. “Hell, it wouldn’t surprise me if you didn’t want me to arrange a press conference for the presentation.”

“Now, you know I couldn’t do that,” Jim told him, shaking his head. “I was told to keep this low-key and avoid publicity at all costs.”

“And look how well you did that,” Gene said, leading Jim away from where Minnie could hear everything.

"Now, I've squared the publicity crap with the company," Gene told him as they got to the rear of the room. "And I've given all the credit for the raids, publicity notwithstanding, to the ICE folks. Toss in the discovery of a massive number of illegal drugs, which have been credited to the DEA, and the company is satisfied.

"It didn't go as I had envisioned," Gene continued. "But this whole publicly noticeable TDA thing was the result of that first gunshot from the two TDA guys behind the motel.

"Had that not happened, I think things would have gone pretty much unnoticed by our media friends," he added.

"All together, I'm happy with the way that side of the operation went," Gene told him. "Especially since we had a relatively unknown guy running that side.

"Having said that, I think Mike did a good job," he continued. "I know he wanted to discuss some things with you and ultimately talked to Butch about the situation, but the bottom line is, he came to the right decision.

"You know there won't be any public recognition for any of us," Gene told him. "But I want you to let both of them know that I'm happy with their first assignment, and if they are interested, opportunities may arise which would suit their particular talents."

"I'll let them know," Jim assured him. "Any reservations about the chainsaw versus scalpel issue I initially had regarding Mike were completely unfounded."

"I may not go that far," Gene said, smiling and shaking his head. "But I will say that he can be restrained with a little guidance. And Butch North is the guy to thank.

"Maybe it's because he and Butch have a close friendship, or because Mike is willing to see things from a different perspective," Gene continued. "However it happened, I think it shows he can temper his initial pulses."

"I agree, sir," Jim replied. "I also hope that we don't have any more situations where we have to make those tough decisions. With a little luck, we should have everyone back here in the next hour. Butch is wrapping up the farmer/parlor operation. Mike says he'll be done in the next few minutes. What's the plan to wrap this up?"

"I'll be talking to the company," Gene told him. "I need to make sure they have all the information, good and bad, and are satisfied with what we've done. Once they decide we've executed our contract as agreed, they will release us."

"What do they think about us turning over the drugs we found at the massage parlors to the DEA?" Jim asked.

"They say that was exactly the right thing to do. We didn't contract for anything to do with drugs, nor did we have the personnel to handle it," Gene assured him. "We were hired to assist in the exportation of illegal immigrants. Everything else was an unintended consequence of executing our contract."

"Or a side benefit to some politician seeking publicity for his upcoming re-election," Jim added. "Just a matter of perspective."

"Isn't everything?" Gene asked as they headed back to the front of the room. "Let's take a farmer's normal breakfast. Ham and eggs. Take the eggs, that's just a pain in the ass for the chicken but all in a day's work. But the ham is a lifetime commitment for the pig."

"Zen Buddhism?" Jim asked, shaking his head.

"No, just an observation," Gene said, smiling. "Now, if you'll resume your role as you were assigned, I'll go see if I can't make arrangements for a small celebration dinner when everyone gets back."

Chapter 61

As the operation wound down and the team leaders along with Butch and Mike began to come back to the hotel, Jim congratulated each of them and headed up to his room to take a quick shower and change clothes before the dinner Gene had arranged.

Coming back down, he was having a Zeigen Bock, looking at the now unmoving screens which depicted where all the action had taken place over the last few hours.

"What's on your mind, Jim?" Gene asked, coming to stand beside him in the vacant conference room.

"Not exactly what we had hoped for, was it?" Jim said.

"Depends on what you hoped for, doesn't it?" Gene asked. "I've seen the reports from most of the cities where we had operations, and I have to admit this one was as good as any of them. Better than most.

"I've also seen the reports from the cities where ICE conducted simultaneous operations, and our success rate, meaning capturing both TDA and Triad members, is better than theirs," Gene told him. "So don't let some preconceived notion of expected outcome lead you to think we've failed.

"As a matter of fact, this operation not only met the goals of Homeland Security, the real force behind the operation, but did it with fewer casualties, both on our side and the targets," Gene continued.

"I still feel as if I didn't do as well as you expected," Jim said, turning to look at the people setting up the tables for the dinner.

"You don't know what I expected," Gene countered. "All I expected was to capture all the targets the company provided. You and your teams did exactly that.

"And as far as the kids go on the Triad side, nobody knew how many there were or where they were," Gene told him. "I'm perfectly happy with what we've accomplished.

"Now, I can't help but think there's more to this than a little adverse publicity you wanted to avoid, or that you wish we could have done something different," Gene said, putting his hand on Jim's shoulder. "I've known you too long not to see something else is wrong. What else is bothering you?"

Jim looked at Gene and confessed, "Marie knows something is going on. She knows I'm up here."

"How do you know that?" Gene asked, surprised that she knew.

"She called and asked if I was here," Jim answered. "I couldn't lie to her, so I told her I was.

"Then she asked if I was on a trip or with you," he continued. "I can only think of one way she knew I was here on one of my supposedly security assessment missions. She had to have seen one of those newscasts where I was videoed."

"I don't see how she could have figured out that any of those were you," Gene argued. "I've seen every one of them that I could find, and if I didn't know you were there, I couldn't have identified you."

"I can't think of any other way," Jim said, shaking his head. "Regardless, she knows."

"What did you tell her?" Gene asked as the tables were covered with tablecloths and settings were laid.

"Not much after I admitted I was here working with you," Jim answered. "She pretty much just hung up."

Gene was silent for a couple of minutes and then asked, "Do you think it would help if I called her?"

"No," Jim answered, shaking his head. "I'll see what she wants tomorrow when I get home. "I just can't figure out what would have upset her even if she figured out I was here."

"It's probably not knowing you are here," Gene replied. "In my experience, it's that she didn't know why. She must think you're hiding something in your life from her.

"Put yourself in her shoes," Gene continued. "If your wife were suddenly seen somewhere, you never thought she'd be and she never told you she was going, wouldn't you wonder why?

"I think once you explain that knowledge of this operation was guarded so as to avoid a repeat of the fiasco in Colorado, she might understand," Gene told him.

"If I'm going to have to tell her every time I accept an assignment from Black Water, am I going to have to explain where I'm going and what I'll be doing?" Jim asked. "I mean, what could I have told her about this? That we're going to go head-to-head with some of the most ruthless men to ever come into this country?

"And what could I tell her about the Chinese crap?" Jim continued. "Then we get into the bigger issue: what do I tell her next time I leave?"

"I don't know the answer," Gene admitted. "But I still believe the biggest issue is that she feels betrayed. Maybe it's time to let her know what we are really involved in. Not the specifics, but more than just telling her you design security systems."

"I guess I think she should just trust me," Jim said. "I never ask where she goes or what she does when I'm on a trip. Why should I? I trust her."

"I don't know what to tell you, Jim," Gene said as people began drifting into the room. "But, if she's important to you, as I believe she is, you need to come to some compromise. Again, if I can help, let me know."

Chapter 62

Gene went to the front of the room as it began to fill up and stood behind the podium as Minnie changed the screen behind him to black. After checking the microphone on the podium, Gene said, "Folks, if you'd please find a seat, I'd like to wrap this up so some of you can get home today."

As chairs scraped across the floor, Gene nodded to Minnie, and she put a map of the United States on the screen. Cities across the entire country had blinking lights covering almost the entire screen.

"The waiters will be here as soon as I tell them it's time to take your orders," Gene announced. "But first, I need everyone to take a seat, any seat, so I can get this over with. If you want to move when I'm done, that's fine. Just be where you want to be when the waiters arrive to take your order."

Seeing most of the people sitting and paying attention to the screen, Gene explained, "The blinking lights you see here represent the cities that we, Black Water, or ICE, the *real* Immigration and Customs Enforcement, hit this morning.

"The goal was to time the raids so as to prevent another Colorado-type fiasco where every single target has disappeared," he continued. "We were tasked with the coordination of our people, and ICE, as well as numerous other federal, state, and city agencies.

"The main reason we were selected for the cities you see now flashing green," he said as Minnie reduced the cities lights to just those Gene was referring to, "was due to one or more branches of the state's governmental offices were known to be reluctant to assist with the Federal Order demanding deportation of any illegal alien who had committed any crime here in the US.

"Let's make one thing clear before we proceed," Gene said as the large number of green lights kept flashing. "Anyone crossing our border without proper documentation, whether that be a visa or another approved document, is technically committing a crime.

"Under the previously mentioned order, the number of potential deportees would have been in the millions," he explained. "There are anywhere between eleven and seventeen million illegal immigrants within the borders of our country this very moment.

"As you can imagine, to track, catch, and deport that number of people would have been an insurmountable task, even with the highest quality of people committed to the task as I see assembled here before me," Gene told them. "Not only do we not have the personnel to conduct such a massive operation, but we would also be fighting our own people in those states that seem to condone the atrocities some of these people commit.

"It was decided by the various agencies of the US government, in conjunction with our folks back at Quantico, that we would only attempt to find and remove those with criminal records," he explained. "Now, that reduces the number to only six hundred and fifty thousand."

Letting them think of the enormity of the task, he continued, "Our part of this was to cover the thirteen states and approximately one hundred and seventy cities that boast of being a sanctuary for illegal immigrants."

Nodding at Minnie for the next view, she replaced the green lights with white flashing lights, and he said, "Here are the cities where ICE or other federal and state officials were conducting simultaneous raids. If you compare their tasking with ours, you'll plainly see that they had almost ten times as many targets.

"They had to cover thirty-seven states and over a thousand cities," Gene explained. "That's an enormous amount of people, and the chances of a leak, such as in Colorado, grow exponentially with every person you add.

"Black Water spent months looking at the entire country and finally decided that with the resources we have, we could only cover those cities where you see the green lights now flashing," he explained. "To compound the matter, some cities are actually considered a sanctuary city even though the state prohibits it, such as here in Indianapolis.

"As you can tell, that makes it very hard to depend on any local assistance, such as you guys ran into at the motel," Gene said. "The state tried to help, but with limited resources and timing issues, they asked the city to lend a hand.

"The city not only did not lend us a hand, instead, they gave us the finger," he reminded them. "That left Mike and his people to deal with an enormous crowd, many of whom actually opposed our efforts.

"So, having tried to demonstrate the enormity of the operation and touching on the coordination to make this seamlessly come together, I want to thank each and every one of you for your dedication and ingenuity in resolving numerous obstacles," Gene said. "Along that line, I'd like to recognize a few of the individuals who were instrumental in making this happen.

"Although I would like to acknowledge everyone who participated," Gene continued, "we wouldn't be through until after breakfast tomorrow. And tonight's dinner is the last meal the company budgeted for.

"Anyway, I'll start with the folks who sat here staring at the screens, keyboards, listening to radio chatter, or keeping track of some of our people who seemed determined to chase across the city," he began.

"Now, don't place any meaning on who is first or the last to be recognized," Gene told them. "I'm just trying to remember all of them, and not having a list in front of me, I'm more or less looking for faces.

"Let's start with Tammie," he said, looking at her near the back of the room. "Please stand up, Tammie."

As she rose, Gene said, "Tammie had the almost impossible job of providing everything you guys in the field needed. Most of which wasn't planned. Nobody thought of the flashbangs, and they aren't usually available at your local Walmart store, except by special order.

"Night vision goggles, another item not found in most Walmart stores," he continued. "Tammie managed to provide everything you asked for. And there's much more. Unplanned buses because we found unplanned children. Ambulances, saws to cut into the back of buildings, the lady is a logistical genius. Thank you. Tammie."

Gene waited until the applause subsided and said, "Now, let's say hello to Amanda. Stand up, please, Amanda."

As she rose from her seat beside Butch, Gene said, "One of the keys to the success of any operation of this magnitude is communications. Not only does every member of each team need to be in contact with one another, but the team leaders also need to be communicating with each other.

"So much of the time, what's happening at one location could potentially happen at another; that's why it's important for the team leaders to know what they are facing at all of the locations," he continued.

"Then, toss in the two leaders of the teams after the TDA and the Triad, the communication could resemble the humming that's heard within a beehive," Gene said.

"Then we have the issue of communication security," he continued. "If Debbie were here, I'd have her stand because she tossed together the system that kept our communication secure from not only our targets, but just as importantly, the media.

"However, the main point is that it took both a knowledge of the various systems of the emergency vehicles, a State Police helicopter that lent a hand, and constant contact with Quantico dealing with the glasses to provide facial recognition," he explained. "And she did it all as calmly and effortlessly as I've ever seen. Thank you, Amanda."

As the applause slowed to sporadic, Gene then said, "The next lady probably did more to help get the dozens of Chinese kids out of what must have been a most horrible life. Or the inevitable life for those recently brought into our country.

"Please stand up, Minzhu," Gene said, turning slightly to look at her. "Folks, say hello to Minnie. She's the reason so many things didn't go sideways with the Triad operation. Her translation once probably saved a gun battle at one of the farms when an unanticipated man appeared with a gun.

"She was also instrumental in translating for a group of Chinese boys and girls who had only been in the country a couple of days and convinced them to let us help them," he continued. "We couldn't have done that without you, Minnie. Thank you."

Again, waiting for the applause to die down, Gene said, "Now, I'd like to introduce a couple of gentlemen who are new to our Black Water family. Mike, would you and Butch, please stand."

As Mike got out of his chair across the table from Butch, and Butch put his hand on Amanda's arm as he rose, Gene said, "These are the two people who tried to herd the cats chasing the mice across the face of the city.

"I can't say enough about how well these two men did their jobs," Gene said. "Now, if you'll give me a second, and Minnie can find the right screen, I'll give you the exact numbers of how well they, and all of you, did."

Nodding at Minnie, he announced, "We took over one hundred and fifty vicious members of Tren de Aragua off the streets of America just here in Indianapolis. Some of these gang members were wanted in their own country of Venezuela for murder, rape, or robbery.

"Not just wanted in Venezuela, most of them had committed the same violent crimes here in the United States," Gene added. "From rape in Georgia, murder in New York City, too many cases of robbery to count, and some unprovoked attacks on innocent civilians who just happened to be walking down the street.

"The numbers are still coming in, but Black Water's operations have removed well over twenty-five thousand of these most vicious animals," Gene told them. "I'm not sure if we'll ever get the exact number of TDA members ICE managed to find. And I'm sure there are several who escaped capture.

"But I know that those who evaded us today will be pursued to the ends of the earth," Gene said. "I'm just proud to say we were a part of this, and I'm proud to have had the chance to work with you guys who broke down their doors.

"Let's take a short break and grab a glass of tea or soft drink before we delve into the Triad side of things," Gene announced as he noticed his phone flashing a light signifying he had a call.

Chapter 63

Gene stepped away from the podium and signaled for Jim to join him. As Jim arrived at his side, he said, "We have a situation."

"What's that?" Jim asked, looking at Gene.

"There was a screwup in Chicago," Gene told him. "There was a major league TDA fellow who we were supposed to pick up at the Cook County jail. This guy was one of the original Tren de Aragua members at one of the prisons in Venezuela.

"He'd been ordered deported, but left and wound up in Chicago," Gene told him. "ICE issued a detaining order when he was arrested on a weapons charge, but the county refused to comply and let him go.

"He was taken into custody again, suspected in the shooting death of a Chicago resident," Gene continued. "Then, he kidnapped a young lady, took her to a house where he was staying, beat the crap out of her and raped her repeatedly over a five-day period.

"When our guys learned he had been turned loose, Quantico was notified and Debbie started running her voice recognition program to see if he had purchased a phone," Gene explained. "She finally located him when he made a couple of calls to one of the phones belonging to one of the TDA members we took out of the apartments.

"My guess is he didn't know about the raids here and was trying to find a place to reconnect with other TDA members," Gene surmised. "Even though he was using a prepaid phone, Debbie managed to bug the phone so we could track him.

"The task was originally given to our folks in Chicago, but by the time Debbie found him, he was halfway down here.

"Quantico asked me to assign someone to intercept this guy since we could get to him faster than Chicago's people," Gene continued. "Now, since this is in Mike's arena, I think he's the guy to take over the capture.

"But it's still your operation and the decision is yours," Gene finished.

"I agree with handing it to Mike," Jim said, nodding. "Is there anything else you can tell me, and Mike, about this guy?"

"Yes," Gene said. "But I'd prefer to talk to both of you, so I don't have to repeat it. And so that if Mike isn't comfortable with the assignment, you can select someone else."

"Just a minute," Jim said. "I'll go get him."

A couple of minutes later, Jim came back with Mike and asked, "Mike, Gene, and I want to know if you're willing to go and try to capture a TDA member who fled Chicago and appears to be headed down our way?"

"When do I need to leave?" Mike asked immediately. "And how should I handle the *arrest* when I catch him?"

"We're getting an Indiana State Police car as we speak," Gene told him. "It should be outside in a few minutes.

"One thing I didn't tell Jim yet, was that the young lady this asshole kidnapped and raped is the sister of one of our people," Gene continued. "He found out about this guy being released and wanted to go after him, but the decision was made at Quantico to use us. Also, the fact of his connection with the lady would obviously come to light, and we don't want him involved.

"Even though I personally wish we could find this guy and hold him until the brother has a chance to confront him, we're afraid that when the target finds everyone missing from the apartment we believe he's headed for, he'll disappear again," Gene finished.

"Are you telling me I need to catch this asshole and hold him until we can notify his brother?" Mike asked.

"Not at all," Gene answered. "All I can tell you is your job is to intercept this guy and make sure he doesn't get away."

"I'd feel better if you took one of your guys with you," Jim said, looking at Mike. "Who do you want?"

"I think I'd like Jon," Mike answered. "I trust the guy and believe he'd be the best one to have in this situation."

"As long as both of you understand the objective," Gene said, "I have no objection to Jon. Just go get him and let's get this going before it's too late. It's about a three-hour drive from Chicago, and he's been on the road for almost thirty minutes.

"That means if you left now, and drove the speed limit, you'd intercept him in about an hour," Gene said as his phone rang.

Looking at the caller ID, he said, "Yes, Debbie. What's the latest?"

Listening to her for a minute, he thanked her and hung up. "That was Debbie," Gene told them. "Our target is still coming down I-65, and there appears to be another man with him from the voices she's heard within the car."

"Go get Jon, if he's who you want," Jim told Mike. "We're running out of time before he hits the city and can get lost a thousand ways."

A couple of minutes later, Mike came back with Jon saying, "Jon's in. What weapons do we have and communication gear?"

"Use the phones from this morning, since Debbie has them in her program," Gene said. "And take the glasses and weapons you were issued.

"As soon as you get the Indy State Police car, give Jim a call and we'll make sure she's tied to both of your phones and can track you and give you any directions you may need if the target leaves I-65.

"If you guys will excuse me, I have to make a quick phone call," Gene said, turning away. "Jim, since these are your guys and now your operation, I expect you to give them any further directions you believe prudent, given the situation."

"Yes, sir," Jim said as Gene walked away.

"What do you think he means about further directions?" Mike asked, looking at Jim. "I thought the General was pretty clear. Catch him and don't let him get away."

Jim looked at Mike, then at Jon, before saying, "Do you remember that little incident you told me about in Vietnam where you were ordered to guard some people and make sure they didn't escape?"

Mike stared at Jim for a second and said," Yeah. I got court-martialed over that *little incident* if you'll remember."

"I remember," Jim said, nodding. "Let's just say this time there's no ambiguity. Catch and detain with extreme prejudice. This guy has used his last get out of jail free card."

Chapter 64

As Mike and Jon were walking out, Gene returned to the podium and said, "If you'll please take a look at the menus on your tables, the waiters will be arriving in a minute to take your orders.

"I apologize for the short break, but I'll try to keep this as short as possible so you guys can get on with your lives," Gene explained as a staff of men and women in white shirts entered the room.

"Let's now look at the Triad operation," he began. "Mr. North, Butch, did a hell of a job. His direction was basically to find all of the Triad people who were running massage parlors employing a bunch of underage girls and some illegal marijuana farms that were using young boys as slave labor.

"Not only did he fulfill his mission, but discovered more than a hundred kidnapped kids who shall be reunited with their families back in China," he continued.

"In addition to that, he uncovered a major drug operation that was operating through these so-called massage parlors, and we captured the major Triad member in this area, who we believe was running the drugs and the child sex operations," Gene informed them. "Among the drugs was over one hundred pounds of fentanyl. One hundred pounds is enough to kill two hundred twenty thousand and eight hundred people.

"Along with the opium, methamphetamines, and other illicit drugs, Butch and his folks have gone above and beyond just finding a member of the Triad," Gene told them. "I don't have the totals from the operations in the other cities, but if they match what we discovered, just the fentanyl alone could cause the death of over thirty-four million of our citizens.

"Once the total drug operations across the country are compiled, this will probably be the largest drug bust in the history of our country," Gene continued. "I'm pretty sure the DEA will make the announcement as soon as they can get all the data from us and their agents in the field.

"As far as I'm concerned, the drugs and saving hundreds of young girls from the horror of sex slavery make this something we can all be proud of," Gene finished.

"We may never know the true extent of the operations against the Triad, but if it saves just the children, I'm happy to have been a small part of this operation," Gene replied. "The only unfortunate thing is that one of our people was severely injured during a kidnapping by one of the madams who ran these despicable parlors. Jewell has been an operative of Black Water for several years.

"The good news is she's due to be released from the hospital tomorrow morning, barring any complications," he continued. "As the only one to sustain a serious injury, I'm eternally grateful. And I credit Butch and his team for that. As I do Mike and his guys for having no serious injuries.

"It goes without saying, as all of you know, your efforts to make all of this possible will never be formally recognized," Gene added as the waiters took orders throughout the room. "But I know. And Jim, who was tasked with overseeing this massive operation, knows. And Black Water knows.

"As far as this operation is concerned," Gene told them, "We've accomplished far more than the company contracted for. And I think the goodwill it has generated will prove beneficial to them. And like all private companies, they depend on making a profit.

"They probably wish they had added a clause where they could expense any additional benefits to Homeland Security," Gene mused. "Let's just say they had a ten percent adjustment for any operation outside the contract. Collecting the TDA or Triad members was all the contract required.

"The drug discovery could have been worth ten percent of the value of the drugs discovered," he told them. "And some value could be placed on saving those kids."

"Ten percent of the negotiated price, which I do not know, would surely be worth a bonus to the men who run Black Water," Gene continued. "You'll notice I didn't say you guys would be included in any bonus.

"But the performance of each and every one of you will be remembered and your names at the top of the list for inclusion when the next operation is staffed," Gene promised. "And since most of you only receive any compensation for the operations where you are involved, that should mean more assignments."

Gene paused and then told them, "I've already had preliminary discussions with the people who make the decisions on issues such as this, and they are looking at adding one day's pay for each of you.

"So, I guess you can consider that a bonus," he explained. "But the biggest thing I think you should take away is the service to our country and communities you have made. Not everything a person does requires a monetary reward. Money is fleeting. Self-satisfaction lasts throughout the rest of your life.

"Now, I'll step down and let you enjoy your meals," Gene finished. "And by the way, Black Water has an open a tab for the rest of the evening, for those who are remaining overnight."

As Gene joined Jim at a table with Tammie and Minnie, he said, "Ladies, if I could have Jim for a couple of minutes, we'll be right back. That is, you'll allow me the pleasure of joining you for the dinner."

Chapter 65

They had almost finished dinner when Gene noticed an incoming call. Excusing himself, he left the table and walked to the front of the room.

"Yes, Debbie," he said. "What's the status of our intercept?"

"Complete," she answered. "The team is returning with an ETA of forty-five minutes. Do you want to speak to them?"

"No," Gene told her. "I'll wait for their return. I'll also let Jim know. Thanks for the call."

Hanging up, Gene returned to the table and said, "I just spoke with someone at Quantico, and they informed me that we are officially finished with this operation.

"What are your plans, Minnie?" Gene asked.

"I'm going to head home, sir," she answered. "I'm ready for a hot bath and a night in my own bed. I don't mean to imply I don't enjoy working with you guys, but this has been a long day."

"I certainly understand," Gene replied. "And I want to thank you again. Helping to get those children to trust Jim was almost a miracle. And I'm sure that just added to the stress of the entire day. Thank you.

"Now, do you need anything? I know you had an open ticket back to Virginia, but have you checked on a flight?" Gene asked as Minnie stood.

"Yes, there's a flight in a little over two hours," she answered. "Just enough time to run to my room and take a quick shower, change clothes, and catch a cab."

"Don't worry about a cab," Gene told her. "I'll have Fabio take you to the airport. He still has a couple of runs tomorrow, so just come down and find him when you're ready."

"Thanks, sir," Minnie said before turning to Jim, saying, "It's been interesting. Not to complain, but trying to keep up with you was sort of difficult. Every time I thought I had a chance to take a breather, you'd be off needing directions somewhere or something."

"My apologies, Minnie," Jim said, standing. "I know I could never have done all the things I needed to do without you, especially when I needed an interpreter. It has been a real pleasure working with you, and I intend to let Debbie know she has a real jewel on her staff. Thank you so very much."

Minnie reached over and shook Tammie's hand, saying, "I don't know how you managed to find all the things these guys decided they needed. I don't think I could have done it."

"It wasn't that bad," Tammie said, shaking her hand. "What I did was nothing compared to keeping track of Jim and the rest of them while managing to bridge the language gap with the Triad. At least I had several breaks where I could just sit back and watch."

"Well, it's been educational working with all of you," Minnie said, turning to leave. "Maybe we'll get to work together again. Just not too soon, please."

"You never know," Gene replied. "If you'll permit me, I'll escort you to the elevator."

"Certainly, sir," Minnie said, waving at Jim and Tammie.

As they walked away, Tammie asked, "What are your plans, Jim? Are you headed back to Texas tonight or staying here until tomorrow?"

"I'll stay here for the night," Jim answered. "Gene sent a plane for Mike, Butch, and I, so I'm guessing he's taking care of the flight home."

"I'm staying tonight, too," Tammie told him. "I could have gotten a flight tonight if I'd known we'd be done this early. But a night on the company in a nice hotel room sounds sort of luxurious right now."

Tammie looked around and asked, "While you're waiting for the General to get back, would you like to go to the lounge and have a drink?"

"Not right now," Jim answered. "I'm still waiting for Mike to get back and see how everything went after his dinner was interrupted."

"Can't you do that from the lounge?" Tammie asked, putting her hand on his arm. "I'm sure he can find us there."

"I don't think that's such a good idea," Jim replied, looking at her hand. "Besides, I'm rather involved with someone right now."

"I don't think there's any need for anyone to know," Tammie said, rubbing his arm. "It's just one night."

As Jim was about to reach for her hand and remove it, Gene walked up saying, "I need to talk to you for a second, Jim, if you'll excuse us, Tammie."

Taking her hand off Jim's arm, she said, "No problem, sir. I was just telling Jim I needed to head up to my room. It's been a long day, and I've got an early flight in the morning."

"If you'd like me to escort you to the elevator, it would be my pleasure, Tammie," Gene said as Jim stood.

"That's all right, sir," she replied, rising from her chair. "I know you need to talk to Jim about something, and I'll just say goodnight to both of you."

"Goodnight, Tammie," Jim said politely. "It's been a pleasure working with you."

Tammie gave Jim a final glance and said, "Goodnight, General."

As she left, Jim looked at Gene and said, "That could have become rather unpleasant if you hadn't shown up."

"I sort of noticed a little flirtatious look in her eye during dinner," Gene said, putting his hand on Jim's shoulder. "That's why I hurried back from escorting Minnie to the elevators. Now, how about we head into the lounge and wait for Mr. Knox to come tell us about his trip north."

Chapter 66

Sitting with Gene and still nursing his glass of Jack Daniel's, Jim asked, "Have you heard from those folks at Homeland Security regarding the success of the operation?"

"Yes, I have," Gene answered. "They are more than satisfied. They never thought we'd accomplish so much in such a short time. I guess their estimates were based on the outcome in Colorado."

"If that was used as the basis for success, we could have quit after the motel," Jim said, looking around at the number of their people taking advantage of the open bar.

"We did have a short discussion about that," Gene told him. "I think some heads are going to roll. Their biggest problem with that is finding who initially spilled the beans. I don't envy that job.

"Reminds me of the old saying during World War II, *Loose lips sink ships*, Gene said. "And it's even more appropriate today with all the texts, emails, and organizations scouring the internet searching for a vulnerable source to exploit.

"Toss in a hacker just a quarter as capable as Debbie, and you have no secrets left," he added. "When an astute marketer can target you because you paused too long on a certain ad or picture on your iPhone, it shows how intrusive these people can be.

"I believe we've also been working with preconceived ideas regarding the capabilities of some of these people, such as Tren de Aragua," he continued. "Just because they come from a country we view as technically behind us doesn't mean they don't have the capability to hack into our systems.

"That would be like saying a seven-year-old kid can't know more about computers than you," he explained. "And we both know that this generation of teenagers is a thousand times more computer savvy than either of us."

"No shit," Jim replied. "Marie's daughters can play a computer like a virtuoso plays the piano, while I'm more akin to playing chopsticks."

"It's a new world out there," Gene said, shaking his head. "I still remember when our enemies sent people to the bars outside a military installation to eavesdrop on the patrons. Now there's some pimple-faced doughnut eater sitting in his mother's basement, trolling the internet.

"I'm grateful we have someone who not only understands the threat, but came from that world," Gene told him.

"I hope you don't have your phone with you," Jim observed, laughing. "I'd hate to see the look on her face hearing you're comparing her to a pimple-faced nerd."

"Unfortunately, I can't leave my phone more than a couple of feet away, given my job," Gene said replied laughing. "Fortunately, she understands how grateful I am to have her on our side.

"By the way," Gene said, looking around the lounge. "Have you seen Butch? I'd like to talk to him and Mike when he gets back."

"No, I haven't," Jim answered. "The last time I saw him was probably thirty minutes or so ago. He and Amanda were leaving the conference room after finishing their dinners.

"I assumed they were coming in here for a drink," he finished, looking around. "I can give him a call if it's important."

"Or I can ask Debbie to locate both of them," Gene replied. "No, I don't think I need to know what's going on with either of them. It's nothing that can't wait until tomorrow.

"I may be getting a little ahead of myself," Gene said, leaning back in his chair. "But since you mentioned Homeland Security, the folks back at Quantico told me the Secretary of Homeland Security is considering a celebration of some sort.

"I'm not sure if it includes all of the heads of the alphabet organizations, ICE, DEA, FBI, or others who participated in today's operations," he continued. "But rumor is that Black Water will be included."

"I guess that means you need to take your tux to the cleaners," Jim told him, chuckling. "I hope you remember the pleats go up ... hence the slang term of *crumberbund*, meaning the pleats catch the crumbs."

"Fortunately, after more than forty years of wearing the monkey suit, I've mastered the technique of dressing appropriately," Gene replied. "I mentioned the celebration mainly because I heard through the grapevine that Black

Water may be doing the same for the leaders of each city's operations. And that includes you, of course.

"So maybe you need to see if yours is still hanging in the closet … or if it still fits," Gene advised. "I think this is the first time I can think of where the company hosts an event following an operation. Maybe because this one was so massive, and the results were better than expected."

"Not to be disrespectful, but I'd just as soon not attend," Jim replied. "I've had enough of that sort of thing while I was still on active duty. There was always a dining-in or dining-out option every year. I think it was more to prove we were civilized and publicly acceptable to the everyday civilian."

"More to instill a modicum of social graces to the average mud-crawling Marine," Gene said, laughing. "At least the dining out gave the ladies a chance to dress up and pretend they had a higher social status than those friends of theirs who had married their high school boyfriends who now ran the family drug store."

Jim started laughing as he looked at the entrance to the lounge where Mike and Jon were entering.

"What's so funny?" Gene asked, moving to see what he was looking at.

"I'm just trying to imagine Mike in a tuxedo," Jim answered, still laughing. "That would be like putting a pinafore on a pig."

Chapter 67

"Have a seat, guys," Gene said as they approached. "We were just discussing you."

'I'm going to run to the bar for a drink," Jon said. "What would you like, Mike?"

"Jack Daniel's, neat," Mike answered. "With a couple of ice cubes."

"Anything for you guys?" Jon asked Jim and Gene.

"Nothing for me," Gene answered. "Don't forget the company has an open tab."

"In that case, make mine a double," Mike told him. "I think after this last thing, I feel like having the company treat me twice as good."

"Our pleasure," Gene told him. "So, how did it go?"

"Pretty easy, actually," Mike told him as Jon returned with their drinks.

"Jon was driving, and we made contact with Debbie as we left the hotel," Mike continued. "She had us heading north on Meridian, just as she told us our target was about eighty miles north of Lafayette, still coming south on I-65.

"She said we were about sixty miles south of Lafayette," he said. "So, I asked her if there was somewhere around Lafayette where we could turn around and let him pass us.

"I figured if we were just sitting beside the road like any normal cop does, looking for speeders, we might be able to sneak up on him," Mike explained. "So, Debbie takes a couple of minutes and tells us that there's an intersection just north of Lafayette where we can exit on State Road 43 and get on the south side of I-65 and enter behind him as he passes.

"So, Jon pushes the speed up to close to a hundred with all the lights flashing and we go flying by eighteen wheelers and Hondas," Mike says, almost laughing. "You should have seen how quickly these people rushed to move into the right lane.

"I know damn well they were doing over eighty before they saw our lights, because all of a sudden, they would slow down and pull behind someone they were about to pass," he told them.

"We keep the speed up just in case the target pulls off before reaching Lafayette," Mike explained. "But luck was with us. Debbie says the target is still coming toward us, and it looks like we'll get to the intercept point fifteen to twenty minutes ahead of them.

"We pass three or four exits leading into Lafayette and are just about to ask Debbie if we missed the exit," Jon says, joining the story. "Then I see a sign saying the exit for River Road is four miles ahead.

"Just as I'm about to ask if that's the same as 43, she tells us we're four miles from the exit and to go left on River Road and pass beneath 65 to a small road called Burnetts Road," he continued. "She said we could turn around there and be ready to get back on 65 behind them, or if they took the exit to Lafayette, we'd be in a position to pull them over before they could make it to the town."

"Jon takes the exit, and we head toward Lafayette on River Road," Mike resumes the story. "Sure as hell, we spot Burnetts less than a quarter mile from 65.

"Jon decides to park on Burnetts so he can see their car when they get to the intersection," Mike explains. "There's a stop sign on Burnetts where it meets River Road, so Jon pulls over just before the sign. We have a good view of the intersection and are just waiting for them to pass heading for Indy.

"Debbie gives us a five-minute warning, and I tell Jon we need to get out of our ICE jackets or if the targets see them, they'll either shoot or try to run," Mike said. "So, there we are standing beside an Indiana State Police car pulling off jackets as cars are passing on River Run.

"Anyway, we turned them inside out, so ICE isn't showing when we put them back on, and we're back in the car when Debbie says the target is only a mile from the intersection," Jon explained. "I'm watching the exit as Mike's rechecking the weapons. All of a sudden, Debbie says, it looks like they are slowing down to take the exit."

"That didn't surprise me," Mike added. "Hell, who knows how much gas they had since Debbie had told us there wasn't shit between Lafayette and Chicago."

"I'm just sitting there ready to turn on the red and blues if they head into Lafayette, which is the only logical move," Jon says. "Less than a minute later, we see a car taking the exit, and Debbie confirms the target just took it. I'm keeping an eye on it as it comes to River Road, and I'll be damned if they don't run the stop sign."

"I don't think they even saw us until they drove right in front of us and Jon hit the flashing lights," Mike told them. "Both of them looked right at us and must have thought we were pulling them over because they ran the stop sign.

"Their brake lights immediately come on and they move to the side of the road," Mike continued. "I tell Jon to take the driver's side, and I'll take the passenger's. Jon has his Glock in his hand, but behind his back, beneath his jacket.

"Since they are both watching him come up beside the car, I step out and hold my Bullpup behind my back as I approach the passenger side," Mike said. "I was slightly behind the guy in the passenger seat, and I had still gone unnoticed."

"That's about when Debbie told us we had a positive ID on the driver as the guy we were after, and the passenger was another TDA guy from the Chicago area," Jon told them. "Mike moved a couple of steps toward the front and tapped on the passenger window with the barrel of his shotgun.

"As the driver turned his head to the right, I brought my pistol to the side of his head," Jon added. "Mike had told me what sort of arrest we were going to make, so I looked up at Mike across the roof of the car, and as he nodded, and pulled the trigger."

"The second I heard Jon's shot, I pulled the trigger also," Mike said. "I know we didn't discuss what to do with any passenger, but I didn't think we could blow somebody's brains across them and expect them not to react. So, I made the decision."

"What did you do with their car?" Gene asked.

"Debbie told us there was a little gravel road about fifty yards from us, and it didn't appear to have any subdivisions from the map she was looking at," Jon answered. "I opened the driver's door and shoved the recently deceased across the console and got in. Mike went back to our car and turned off the flashing lights and followed me to the turnoff.

"The road made a sharp left turn about two hundred feet from River Road, so we parked their car at the end of the road, another hundred feet or so from the turn," Mike added. "It wasn't perfect, but we didn't have too many options at that point. Hell, I think we were lucky nobody came by when we were standing there with our weapons pointing in the windows."

"I'll make a call to the Superintendent of the State Police," Gene told them. "If he can get someone out there early enough, maybe they can just tow the car somewhere and run it through one of those crushers and ship it off to be made into one of our American cars made in China.

"If it's found by one of the locals, I'll at least have given him a heads up before it hits the news," Gene added. "And with all the press about the operations here and in Chicago, I have no doubt some enterprising reporter will make the connection.

"That's still their problem," Gene finished. "We'll return the car we borrowed and thank them for helping us."

"About the car," Jon said. "When I got out of the target's car. I had unfortunately sat in some rather unpleasant gooey material. I'm afraid we didn't clean their seats before we brought it back."

Chapter 68

The next morning, Jim came downstairs and met Gene in the hotel restaurant. "How's the after-action clean-up going?" Jim asked, sitting across from him and picking up the menu.

"As normal as it ever is," Gene answered, looking up from the newspaper he was reading. "Lot of ink wasted telling all the good citizens that some very bad people have been sent home."

"I've been thinking about that," Jim said as a waiter approached the table. "How long do you think it's going to take to get that many people back to wherever they came from?"

"Longer than it took us to catch them," Gene answered as Jim ordered eggs Benedict, orange juice, and black coffee for breakfast. "You know as well as I that the airline industry can't handle the deportations.

"How many passengers want to be on a flight with deportees?" Gene asked. "At least thirty percent of the people don't favor deporting even the criminals who are here illegally. At best, seventy percent don't want to deport any of the other groups who are technically here illegally.

"And then there's the danger to the crew. There was an incident involving deporting Haitian illegal immigrants when their plane landed in Port-au-Prince," he explained. "They attacked another plane carrying families. These flights were contracted through ICE.

"They attacked the pilots and three ICE officers," he told him. "Given the high risk of the people we've taken, including those ICE found, that limits the number of flights available.

"Just think of how many 737s it would take to move six hundred and fifty thousand people," Gene said as his breakfast was delivered. "A 737 carries what, one hundred and thirty people? That's five thousand flights.

"Last time I saw any numbers, ICE only has twelve aircraft," he continued. "That means there will be lots of charter or contract operations, along with military aircraft such as the C-17.

"Just be glad you're not one of the pilots," Gene finished as Jim's breakfast was set on the table.

"I have enough trouble with some of the passengers who are just normal people," Jim told him. "You wouldn't believe how many people don't understand that keeping your seatbelt on while taxiing is because of sudden stops or sharp turns.

"And don't even mention those folks who want a blanket or a pillow," Jim said, shaking his head. "I feel sorry for the ladies in the back who have to deal with them. I damn sure wouldn't want a bunch of TDA animals back there."

"Morning, guys," Mike said, walking up to the table. "Mind if I join you?"

"Have a seat, Mike," Gene answered. "I guess you can tell we didn't wait for you."

"I'm used to being the last dog to the bowl," Mike told him as the waiter approached. "Satisfied to eat the scraps."

"Have you talked to Butch since last night?" Jim asked as he took a drink of the orange juice.

"No," Mike answered as he ordered steak and eggs over easy with coffee.

Handing the menu to the waiter, he continued, "None of my business, but I called his room last night before I went to bed, and again this morning before I came down here."

Jim looked at Gene, and then back at Mike, saying, "I'll bet he was in the shower when you called last night and probably went for an early jog this morning."

"That's possible," Mike responded. "It's just that I saw Amanda get in the elevator with him when I left you guys last night."

"It's not unusual for more than one person to need to use the elevators at the same time," Gene explained. "Hell, I've had to share an elevator with six or seven people sometimes. Not unusual at all."

"They probably got off at different floors, anyway," Jim said, smiling slightly.

"Even if they didn't," Gene said. "Several of our people are on the same floor. Matter of fact, all of us are on only two floors."

"It's probably all a coincidence," Mike said, nodding. "And he probably had his arm around her because she had one too many beverages."

"Undoubtedly," Jim said as they saw Butch and Amanda coming into the restaurant. "I've heard he's a very considerate gentleman."

"Good morning," Butch said as he came to the table with Amanda. "How's the breakfast menu here?"

"A little limited," Jim answered. "I was hoping for huevos rancheros, but it wasn't on the menu. Would you guys like to join us?"

"That's all right," Butch said. "Looks like you're already eating, and that table sits only four. I'd join you, but I ran into Amanda coming down on the elevator and told her I'd join her."

"I understand," Gene said, wiping his chin to hide his smile. "But we need to be at the airport in about an hour for the trip back to Texas. Fabio will be out front in forty-five minutes or so."

"I can make that," Butch said looking at Amanda. "I'm just having coffee and a bagel."

As they headed for another table, Mike whispered, "I've never heard it called that. Lots of other names, but never a bagel."

Chapter 69

After landing at Dallas Love Field, Gene thanked everyone and said he had to make a couple of other stops before heading back to Quantico.

Driving back to his house so Mike could get his car, and they could head home, Jim asked, "What do you guys have planned for the next few days?"

"I'm going to drop Butch off and then go home and get rather shifazed," Mike answered. "This has been more fun than I've had since I tried to get that lady at the Rodeo Exchange to take me home."

"What the hell is shifazed?" Butch asked.

"That's French for shit-faced," Mike answered. "You should get some culture in your life."

"How'd that work out?" Jim asked.

"I was beat out by a group of firefighters," Mike said, laughing. "Guess she thought they had bigger hoses."

"Or they weren't looking through whiskey goggles," Butch argued.

"Didn't matter," Mike said. "I probably saved myself from another disastrous divorce in a year or so. You'd think two previous mistakes would teach me not to grab the first shiny thing I see when I walk into a bar."

"Ah, the problem with the three D's," Butch said. "Drunk, dark, and distance. The best friends a lady has when out on the town."

"Don't forget the whiskey goggles," Jim added. "Surprising how good some lady looks when you're wearing them. Then daylight comes and you're looking at the wicked witch of the west."

"But the hours between are what keep us looking for true love," Mike said. "I think I've stepped over a thousand perfectly good women to find the absolutely wrong one standing alone in the darkest corner of the bar."

"And yet you keep trying," Jim said as they neared his house. "Shows superb determination."

"You have to admire a man who knows what he wants," Butch said. "Sets high standards and fails to meet them repeatedly."

"I do have my standards," Mike said as they pulled into Jim's driveway. "But I also have waiver authority."

"How about you call me when you're in the mood to go searching for that higher standard lady?" Butch asked as they got out of Jim's car.

"You're on," Mike said, taking his bag from the seat beside him and getting out. "Just give me a few days. I need some time to relive the last couple of days.

"Private jets, lots of action, and not to mention, a couple of very interesting ladies," he continued.

"What do you have planned, Jim?" Butch asked tossing his bag in Mike's car.

"Mending fences, I guess," Jim answered.

"Is there someone you're keeping secret?" Mike asked, standing beside his car.

"Not really a secret," Jim answered. "And, if I can straighten things out, I'll have you guys over for a post op drink to meet her. Maybe you can see what standard you should try to attain."

"Sounds like a good plan," Butch said nodding. "In the meantime, are we going to hear from the General if there's another operation for us?"

"That I can't say," Jim answered. "I go months before I hear from him sometimes. Black Water has stuff going on around the world all the time.

"But, Muddy Water, the subsidiary we were working for this time, only works within the country," he explained. "I'd say they will contact you when you least expect it. But you can always decline if it's not convenient or not something you're interested in."

"Guess that's sort of like when I give a lady my number," Mike observed. "Waiting to see."

"That's the best I can tell you, guys," Jim told them. "But given the success we had, I'd bet you'll hear from them sooner or later. And I have to say it's been a pleasure working with you.

"It may not mean much, but I'm glad to have had you guys making my job easy," he told them. "Thanks."

Shaking Jim's hand, Mike said, "I'm glad you recommended me. If there's ever anything I can do for you, just ask."

"Me, too," Butch told him. "Even if the company doesn't call, let's keep in touch."

"I'll definitely do that," Jim said as they turned to Mike's car. "And I'm sure our paths will cross again. Until then, you have my number. Call if there's anything I can do for you."

As they got in the car and left, Jim took his bag into the house and tossed it on the living room floor. Walking into the kitchen, he took a bottle of Jack Daniel's from the cabinet and a glass before walking outside to think of what he would hear from Marie when he called to tell her he was home.

Chapter 70

"Hey," Jim said when she answered his call. "What are you doing today?"

"I promised Dad I'd come help at the restaurant," she told him. "Mom's not feeling well, and Dad doesn't want her around the food."

"I can understand that," Jim said. "I just thought I'd call and let you know I'm home."

"Okay," she answered. "Why don't you come to the restaurant about three or four this afternoon. We're never too busy at that time, and we can have a talk.

"I'll even save the table you like so well in the back," she finished.

"I'll be there," Jim replied. "Is there anything you want to talk about right now?"

"No," she answered. "I'll wait because I want to show you some stuff."

"What stuff?" Jim asked.

"Not right now," Marie answered. "I've got customers coming in and I need to get back to work.

"You'll see when you get here," she continued before hanging up.

'Well, that didn't go so well,' Jim thought as he took his empty glass back to the kitchen.

Tossing the dirty clothes from his bag into the hamper in the bathroom, he looked to make sure he had a clean uniform for the flight he was scheduled to take tomorrow morning.

Deciding to do a load of laundry while waiting to go see Marie, he had just started the washing machine when his phone rang.

"Captain Lashley, this is crew scheduling," the voice said. "We need your trip tomorrow for training. Is that acceptable?"

"That's fine," Jim answered. "Thanks for the early call."

"Sorry it's this close to your trip," the caller apologized. "I just got notified an hour ago that a Captain upgrade was delayed due to a weather issue in Chicago, and your trip fit their requirements."

"Not a problem," Jim replied. "I don't care if you don't let me know until just before I walk out of the house on my way to the airport.

"A day when I get paid to stay home is a great day," Jim continued, smiling to himself.

"I wish we had that opportunity," the caller replied. "You have a good day, sir."

"You, too," Jim responded before hanging up.

A few minutes later, his phone rang again, and checking the caller ID, he answered, "I'm not surprised to hear from you about now."

"Oh," Gene said. "Why would you be expecting me to call?"

"As I'm sure you're aware, I was just removed from my trip tomorrow," Jim answered.

"That's certainly a coincidence," Gene told him. "Because I've just been informed that we need you to come to Quantico in two days."

"Yeah, let's call it a coincidence," Jim replied, laughing. "So many coincidences with the company. What is it this time?"

"If you'll remember, I told you the company might be planning something," Gene told him. "Well, it's been planned for two days from now.

"I'll send the plane down that morning and have a room for you at one of the hotels," Gene continued. "We've booked over one hundred and fifty rooms, so don't be surprised if you run into some of the folks from yesterday.

"Oh, don't worry about the tux," Gene added. "A sport coat and your usual Wranglers will fit in just fine."

"I believe I told you I'd just as soon not attend," Jim replied, shaking his head. "Or isn't that an option?"

"I'm afraid it's not optional," Gene answered. "Unless you don't want to continue employment with the company."

"That may be up for discussion this afternoon anyway," Jim told him. "I'm meeting Marie at the restaurant in a couple of hours, and I have no idea what's on her mind."

"Do you really think she'd throw in the towel because you didn't tell her about this operation?" Gene asked.

"Hell, I don't know," Jim replied. "I called, as requested, and was more or less put on hold.

"She said she had something to show me," he continued. "I'm guessing it's the videos from Indy, but I still don't know how she could have identified me.

"Unless someone up there videoed me without my helmet and stuff," he said. "You don't think someone at the hotel made any videos, do you?"

"I really doubt that," Gene answered. "What would be the point? Who would risk their job for that?

"Now, there's a slight chance that you were seen entering the hotel or leaving it without your helmet and face shield," he continued. "Possibly someone making a video of another person or people, and you just happened to be in the background."

"And a snowball stands a chance in hell," Jim argued. "No, it has to be from one of those videos from the car wreck."

"You're probably right," Gene agreed. "But I don't know how she could identity you from any of them. I'd be surprised if Debbie's facial recognition program could do it."

"Well, I guess I'll find out in a couple of hours," Jim repeated. "I guess I'll let you know about that little celebration thing when I get home from the restaurant."

"Okay," Gene said. "I have a few other calls to make to tell the other cities' leaders to go find a clean shirt and some starched jeans."

"Before you go, are Butch and Mike invited?" Jim asked.

"No," Gene answered, "but I don't think you should be surprised to see Butch on the flight coming up here."

"Really?" Jim asked.

"Yep," Gene responded. "Seems that one of the other people who received an invitation added him to her RSVP."

"Amanda," Jim said.

"You have a sixth sense, my friend," Gene said, laughing. "Give me a call when you get home."

Chapter 71

At three o'clock, Jim headed for Siciliano's, a little apprehensive about meeting Marie. Not only was this the first serious situation between them since they first met, but the pressure of having her father there made him wonder if he should have said he couldn't make it this evening.

Then there was the issue of having to go to Quantico in two days. If she had a problem regarding his association with either Gene or Black Water, there would be a tough decision to make.

As much as he wanted to make a life with her, and really enjoyed her parents and the twins, he didn't want to relinquish his relationship with Gene. Nor did he want to stop working with Black Water.

He saw that part of his life as a continuation of his time in the Marines and the service to the country. Regardless of the assignment, he always saw the reason behind it. And like this last one, he saw it as necessary to protect and defend his nation.

As the restaurant came into view, the urge to turn around and go home was overwhelming. But, as with most things in his life, he knew he had to face it, no matter the consequences.

Marie met him as he walked in and asked, "Do you want anything to drink?"

"No, thanks," Jim answered as she led him to the back of the restaurant.

Taking a chair across from where she pulled out a chair, he asked, "So what is it you wanted to show me?"

Marie picked up some photographs that had been on the table when they sat and said, "Can you tell me what this is?"

Jim flipped through the photos and asked, "Yeah, they're pictures of me while I was in Indianapolis."

"I know that," Marie replied, taking out her phone and thumbing through it until she found what she was looking for and passed it to Jim.

As Jim watched the scene where he confronted the obstinate guy at the scene of the car wreck, she said, "I never would have recognized you in those pictures.

When I heard you tell the guy to get the hell back in his car, I was pretty sure it was you," she continued. "I know the way you talk and sound.

"Then later, when you got in his face and called him *scooter*, I knew it was you," she said. "I've never heard anyone, but you, call anyone scooter."

"Okay," Jim said, handing her phone back. "I've already told you I was there, so what's the problem?"

"The problem, as you put it, is that you never told me you were working for the Immigration people, ICE," she said. "How long have you been working for them?"

Jim sat back thinking of how to explain what the operation in Indy involved, and finally said, "I don't work for ICE. I was there working with Gene, in conjunction with other federal agencies.

"I don't know if you saw what happened in Colorado when ICE raided an apartment looking for that gang from Venezuela, the Tren de Aragua," Jim said, leaning forward.

"After that fiasco, some branch of the government, I believe Homeland Security, contracted with Gene's company, Black Water, to help launch a one-day operation to find and capture all of the known members across the country," he explained.

"After the Colorado thing, there was to be no discussion of the raids unless the people were part of the operation," he told her. "My part was supposed to be directing the teams who were actually involved in the raids.

"Unfortunately, one of the targets escaped with a hostage," Jim said. "Now, another part of the raids was against the Chinese Triad. The target that escaped was one of that group.

"Along with the hostage, who was a member of Black Water, she took four of the little girls who lived in the apartment with her and other girls," he continued. "The lady, if you want to call her that, was running a massage parlor where the little girls were forced to work."

"Are you telling me that those kids were working in a massage parlor where they were expected to do … you know?" Marie asked, staring at Jim.

"Yes, that's exactly what was going on," Jim answered. "And they also had illegal marijuana farms where the boys who were brought in were made to work.

"The reason for the ICE jackets we all wore was to make sure if some local law agency came in contact with us, they would assume we were with Immigration," Jim explained.

"And, to give us some appearance of authority and disguise our people."

"What happened to those kids?" Marie asked. "Were they all right after the wreck? What's going to happen to them now?"

"I don't know about their condition after the wreck, but one of the EMTs said there was nothing serious," Jim answered. "As far as what's going to happen to them, I've been told that they will be returned to their families back in China."

"I can't believe people would use little girls that way," Marie said, shaking her head. "Those girls I saw couldn't have been more than twelve or thirteen. That's so disgusting."

"I know," Jim said, now happy about his decision to come face her. "This was the most difficult mission I've ever been given by Black Water. But I'd gladly do it again if it stopped the kidnapping and sexual abuse of those kids."

"What's going to happen to those people, like the lady you said took one of your guys hostage?" she asked.

"They will be deported," Jim answered. "Except that lady. She died in the accident. Maybe they shipped her body back, but I don't know.

"We just took everyone, TDA, Triad, children, whoever, to a location where the federal guys made whatever arrangements to return them," Jim finished.

Just as Marie was reaching across the table to take Jim's hand, her father walked in and announced, "Look who else came to visit my poor little establishment."

Chapter 72

"Good afternoon, everyone," Gene said, coming in with Tony. "Hope I'm not interrupting anything."

"Not at all, Gene," Marie said, standing. "We were just talking about you."

"I hope it was something reasonably good," he answered, giving her a hug.

"Of course," she replied, motioning for him to take a seat.

"May I bring a bottle of wine and some appetizers?" Tony asked as Gene took his seat.

"It's pretty close to dinner time for me," Gene said, looking at Jim and Marie. "I'm still on East Coast time, so I'm going to have something a little bit more along those lines."

"What did you have in mind?" Tony asked. "Or do you need a menu?"

"I've heard your lasagna was just fine," Gene said, grinning. "At least I believe that's what Jim told me."

"I believe we've had this discussion before, my friend," Tony said, looking at Jim and shaking his head. "But I'll bring my best-in-the-Metroplex lasagna. And I'll bring a little something for you while waiting for your meal.

"Would anyone else like to order?" Tony asked, looking at Jim.

"Shrimp scampi sounds just fine for me," Jim said, trying to stifle a laugh.

Tony put his hands on his hips and said, "I'm starting to wonder if I even like you, sir. But I guess since my daughter seems to think you're salvageable, I'll overlook your obvious inability to discern succulent dishes such as I prepare."

"Then I guess I should say the shrimp scampi sounds succulent to me," Jim replied, laughing.

"Anything for you, my dear?" Tony asked Marie.

"Lasagna sounds just fine for me, too," she answered as they all laughed.

"Heathens," Tony said, shaking his head as he turned to leave. "Now even my daughter is infected with tastelessness."

"I don't know if Jim told you or not, but he's been invited to a small celebration at Quantico in a couple of days," Gene said, looking at both of them as Tony left.

"No, we just talked about where he was for the last couple of days," she answered.

Looking at Jim, she asked, "Don't you have a trip starting tomorrow?"

"I got replaced," Jim answered. "Something to do with needing it for some training."

"Back to the celebration," Gene said as he pulled an envelope from the inside of his jacket. "I've been asked to deliver this formal invitation personally."

Jim looked at him and took the envelope, which had his name printed in script on the front.

Opening the envelope, he pulled out a single folded sheet with an embossed seal of the United States. Quickly scanning the page, he looked back at Gene and asked, "Does this have anything to do with the event Homeland Security is having?"

"No," Gene answered, shaking his head. "That one will be tomorrow evening, which is why I'm in sort of a rush today.

"That one is just for the heads of the agencies that participated in the deportation operation," he explained. "I'm going as the representative of Black Water.

"This one is just for the Black Water people who were involved," Gene clarified. "And it's mainly for those people who led the operations at each of the cities we were assigned to cover.

"Jim has expressed some reluctance to attend," Gene said, looking at Marie. "But, considering who is coming, I believe it's important not just to our company, but to Jim and the others who were instrumental in the biggest illegal immigrant deportation in the history of our country.

"And it was more than just a deportation of illegal immigrants, it was a cleansing of some of the most vicious and violent criminals found on the face of the earth," he added, glancing at Jim.

"So, as much as he'd like to spend some time with you on his days off, I want him to attend," Gene finished.

"Who is the person you said is coming?" Marie asked.

"If you noticed the seal on the letter, you could guess it's from someone fairly high in the government," Gene told her. "Since it's involving the itinerary of a high-ranking official, it's classified.

"However, since I know neither you nor Jim will let anyone know, it's the newly appointed Border Czar, Ted Hastings," Gene said. "He's coming specifically to thank the people who made this operation so successful.

"And I'm personally extending the invitation to you, Marie," he concluded. "There will be a small celebration for our people after Mr. Hastings makes his presentation. I've already told Jim there are reservations at the hotel, and I'm sending my personal jet to bring him, and you, if you want to come."

Both Jim and Gene sat quietly looking at Marie until she finally said, "It would be my pleasure.

"Now, what sort of dress should I wear? Is it formal? Will there be other women there? Is there another woman who can call and tell me what she's wearing?" Marie asked in rapid succession.

"What time is the plane? I've got to go shopping for something appropriate," she continued. "Shoes, purse, oh, I'm going to need a new necklace and some other jewelry.

"I almost forgot, I need to make an appointment to get my hair done. Do you think you can have the plane come later in the evening so I can get all of that done?"

Gene and Jim looked at each other and slowly shook their heads.

"Marie," Jim said. "The plane isn't coming to pick us up until the day after tomorrow. You can shop all day tomorrow, but I'm just wearing jeans, a starched white long-sleeved shirt, and a sport coat."

"And the plane won't be at Love Field until eleven o'clock, so you have a couple of hours that morning," Gene added. "Tell you what. I'll have Amanda give you a call, and she can give you an idea of what most ladies wear at one of the informal affairs."

Chapter 73

The next morning, Jim woke up at Marie's house, hearing her on the phone talking to someone about what sort of dress to wear, high heels or flats, and a laundry list of other items.

Walking into the kitchen where she was sitting at the small bar separating the kitchen from the dining room, Jim poured himself a cup of coffee and took the stool next to her.

Looking down at the pad where she was listing everything she thought she needed to do before the next morning's flight, he waited patiently until she said, "Sorry, Amanda, I've got to go. Jim's finally awake, so I can send him home and start on this list of things I need to get.

"And thank you so much," Marie said, holding up a finger to Jim, signifying she needed another minute. "I'm looking forward to meeting you tomorrow evening. You've been such a help; I was going nuts trying to figure out what I needed to wear.

"I'm thinking a simple black dress, off the shoulder on one side, and medium pumps, black of course, and a matching handbag with a silver chain," she continued. "This is going to be so much fun."

Finally hearing her tell Amanda goodbye, Jim said, "I guess I'd better head home so I can make sure I have a pair of starched Wranglers in the closet. Oh, I need to see if I need to polish my brown boots that match, sort of, my brown belt."

"You guys, you don't have a clue as to how important it is to a lady to dress appropriately," Marie told him, standing to refill her coffee cup. "But you're right, I need to get started on this list if I'm to have everything ready for tomorrow.

"Oh, Amanda said she's invited Butch North," Marie said coming back. "I guess he's flying with us since she said he lives out here somewhere. Do you know him?"

"Yeah, I'd met him a couple of years ago. He flies for American, also," Jim answered. "He worked with me on the Chinese Triad problem in Indy. Nice guy."

"Amanda is such a sweet lady," Marie said, reviewing her list. "Maybe she'll come down and we can visit with her and Butch. Maybe have a little party at the restaurant."

"Don't you think you're getting a little ahead of yourself?" Jim asked. "I'm not positive, but I'm pretty sure they just met during the operation in Indianapolis. He's going up to the celebration as her guest. But I think it's a little early to be planning their engagement party."

"I'm not planning an engagement party," she argued. "I'm just thinking of a way to thank her for all the help she's been."

"Okay," Jim said, getting up and dumping the remains of his coffee in the sink. "I know you want to get started on your shopping spree, and I need to get home and do some laundry or something equally as exciting."

"Do you want me to come over and spend the night this evening?" she asked as she followed Jim to the door.

"That's up to you, but you might want to stay here in case you need to go shopping for some last-minute things," Jim told her. "As much as I would like for you to spend the night, I think it would be better if you came over just before we need to be at the airport.

"If you want me to come get you, just let me know," Jim told her as they walked out to his pickup. "Just remember that we need to be at the airport no later than eleven o'clock."

"Since it's a private jet, can't we set the departure time to fit our schedule?" she asked. "I mean, what if I'm running a little late?"

"It is a private jet," Jim agreed, opening the door to his truck. "But it's not my private jet. The flight is probably scheduled to ensure we can arrive at the hotel and then proceed to the celebration venue.

"So, the answer is no," Jim told her. "Just treat it the same as you would if you were flying on American or any other airline. At least you don't have to be at the airport two hours before the flight in this case.

"And, no security bag checks or crap," Jim added as he started the engine. "Now, if you decide to come over tonight, just let me know and I'll make dinner."

"You're probably right," she said. "I'll just plan on being at your house around ten o'clock for the flight. Anyway, I've got so much to do, so I'll give you a call this evening and tell you all about it."

"Sounds good to me," Jim replied. "I'll talk to you then."

As Jim was heading home, his phone rang. Checking the caller ID, he answered, "Good morning, General."

"Good morning," Gene said. "I guess the issue with Marie is resolved in your favor."

"I guess," Jim said as he reached the 635 loop around Dallas. "I don't think I've ever seen an attitude go from pissed to ecstatic so quick."

"It's amazing how that happens," Gene replied. "I'm sure most of it was whatever explanation you gave her. But the chance to dress up and attend something as unusual as this can be a mood changer pretty quickly."

"That and the chance to go shopping," Jim added, laughing. "Anyway, the issue was resolved without having to disclose some of the stuff we contract to do that could be a little more difficult to explain."

"You're right about that," Gene said. "Anyway, I don't think I told you where the event will be. We thought about having it at the hotel, but there's the issue of security, and we didn't get the promptest notice that the Border Czar was coming.

"So, it will be here at Black Water headquarters," he explained. "I'll have a limo waiting at the airport. I figured what the hell, if this puts a smile on Marie's face, let's go a step further and treat her to something special. Could gain you some bonus points for the next time you screw up."

"I don't care how many *bonus points* I can accumulate," Jim said, pulling into his driveway. "One *aw shit* erases all of them. And as you know, I can step in it without knowing I've stepped in it."

"I understand," Gene agreed. "Anyway, I have a lot to take care of between now and then. I'll see you tomorrow afternoon."

Chapter 74

The next morning, Marie called to say she was almost ready to go, and they could stop for an early lunch if he wanted. Saying that would be fine, he hung up and put his fresh jeans, shirt, and sports coat in a hang-up bag.

Heading back into the kitchen for another cup of coffee, his phone rang again. Not recognizing the number, he just said, "Hello."

"Jim," the caller said. "This is Butch. Butch North. Did I catch you at a bad time?"

"No, just having a cup of coffee," Jim answered. "What can I help you with?"

"I was just wondering if it would be all right with you if I came by and left my car at your house?" Butch asked. "I can be there in about an hour. Will that be okay?"

"No problem," Jim answered. "Do you need directions?"

"I don't think so," Butch said. "I've got this map thing on my phone that's supposed to be pretty accurate."

"Just give me a call when you think you're close, and I can make sure you're in the right neighborhood if you have any trouble," Jim told him.

"By the way, Marie, the lady I've been seeing, is coming with me," Jim explained. "And we're going to have a quick lunch before going to the airport. You're welcome to join us if you'd like."

"That sounds great," Butch replied. "What time should I be at your house?"

"I guess about nine o'clock," Jim answered. "That should be plenty of time to stop at Denny's and eat and be at the airport by eleven."

"I'll be there," Butch told him. "By the way, Amanda called me last night and told me she had been talking to Marie. At least Marie will have someone there to talk to when we get there."

"I know," Jim said. "Just be ready for twenty questions on the flight to Quantico. And I shouldn't have to remind you; she's not part of the organization as Amanda is. So be careful of what you say."

"Got it," Butch told him. "Anyway, I'm tossing my bag in the truck, and I'll see you in an hour."

As his phone rang again, Jim muttered, 'What the hell is it this time?' as he looked at the caller's ID.

"Yes, sir," Jim said, answering. "What can I do for you this morning?"

"I just wanted to tell you that you'll be stopping in Memphis and picking up Tammie," Gene told him. "Her name was included sort of at the last minute, and I didn't get a chance to get her a commercial flight.

"And she's been briefed that Marie isn't part of the company, so she doesn't say something Marie doesn't need to know," Gene finished.

"I guess you didn't think about the little, as you said, *flirtation* thing," Jim replied. "And now you put her on the same plane with Marie and me?"

"You're a big boy," Gene told him, laughing. "And I don't think Tammie will do anything that could be considered inappropriate, knowing Marie's the lady you've been seeing."

"I just got out of one heated issue, and now you start stirring the pot," Jim told him. "I'm not sure if you're on my side or not."

"I'm always on your side," Gene said. "But I think in this case, you're worrying about a perceived problem before there's a problem. Hell, you were probably going to run into her at the celebration anyway.

"Now, I've still got a lot to do before you get here," Gene finished. "I hate these high-level politically inspired gatherings. Too many little protocol issues."

"I didn't think this was political," Jim said. "I thought it was just a thank you party from an appointee, since a Czar isn't a cabinet position."

"No, Ted isn't a member of the cabinet," Gene confirmed. "But at this level, everything has political ramifications. Now, I've really got to go. Give me a call when you get to the hotel."

A few minutes later, the doorbell rang, and Jim went to answer it. Seeing Marie's car at the curb, he opened the door, asking, "Do you want me to put your suitcase in the truck?"

"Sure," she said. "But there are two suitcases."

"Not a problem," Jim said, walking down to her car. "This airplane holds about fourteen passengers.

"Oh, we're stopping in Memphis to pick up a lady who's going with us," Jim said, setting her suitcases in the rear seat.

"Oh, I guess that's all right," Marie said. "Do you know her?"

"Sort of," Jim answered as Butch drove up in his old pickup. "She was part of the team that stayed at the hotel. Her job was making sure everyone had whatever they needed."

"Hey, Jim," Butch said walking up the driveway with a suitcase. "Thanks for letting me ride with you to the airport. I've never been to Love Field, and I hate driving in Dallas."

"My pleasure," Jim said, taking his suitcase. "Butch, this is Marie. Marie, meet Butch."

"Pleased to meet you, ma'am," Butch said. "I sort of know about you through Amanda."

"Very nice to meet you, too," Marie said. "Amanda was so nice helping me decide what I should wear. This is sort of my first high society fling."

"I wouldn't call it high society," Jim said after putting Butch's suitcase in the pickup. Last night's deal was more of a high society. Black tie, evening gowns, ladies wearing gloves and stiletto heels, and probably enough sparkling diamonds worth more than it takes to run a small nation for a couple of years.

"You guys go ahead and jump in," Jim told them. "I'm just going to grab my bag, and we'll head for Denny's."

Chapter 75

After landing at Quantico, they were met by Fabio at the airport. "Fabio," Jim said as he saw him standing by the limo. "I didn't think I'd see you again so soon."

"I got the privilege of driving you and your friends to the hotel," he said, shaking Jim's hand. "Can I give you a hand with the luggage?"

"Sure," Jim said. "By the way, this is Marie, a very special friend of mine."

"Marie, so glad to meet you," Fabio said as he shook her hand. "I had the honor of being Jim's driver a couple of days ago. Most interesting."

"Nice to meet you, Fabio," Marie said. "I can't imagine just being his driver would be that interesting, but I'm glad you thought so."

"Possibly not what I envisioned, but it was truly an experience," he told her as he started putting their bags in the trunk.

"Am I your only assignment this afternoon?" Jim asked as the last bag was put in.

"Oh, no," Fabio answered as he shut the trunk. "I'll be running back and forth a couple of times before I'm through picking everyone up.

"You guys are just my first trip," he told them as he opened the door for Marie to get in.

"Well, I guess you'll be busy for the next couple of hours," Jim told him as he got in. "But it's good to see you again."

After arriving at the hotel, Jim and the others went to the desk to sign in. As he did, the clerk handed him a package and said, "Sir, there are some forms for you and the others that need to be filled out and returned to the front desk as soon as possible. If you have any questions, please call the number listed on them."

"Thanks," Jim said as the clerk handed him the key card for his room.

As each of them signed in, they were also given a package along with their keys.

Heading for the elevator, Jim said, "I guess we'll see all of you at the headquarters, but if you have any problems, I'm in room 325."

After arriving at his room, Jim opened the package and saw a proposed time frame, along with a form requesting their choice of meals: chicken or steak, and a selection of vegetables.

"Steak or chicken?" Jim asked as Marie started unpacking.

"Chicken," she said. "What are the veggies?"

"Mashed potatoes, green beans almondine, or corn," Jim answered. "Pick two."

"Potatoes and green beans," she answered. "I'm going to jump in the shower. Is there anything else they want to know?"

"Red or white wine?" Jim told her as he filled out the form, which included the table number and seat location.

"White," she said as she hung her dress in the closet.

"Got it," Jim told her as he finished filling out the form. "I'll take this down to the desk while you shower, and then I'll come back and take a quick shower. We still have almost an hour before we need to be downstairs for the pick-up.

As Jim finished showering, Marie said, "You know, that lady we picked up in Memphis is interested in you."

"Really?" Jim asked, taking his jeans from his hang-up bag.

"Really," she replied as she looked in the mirror and fastened her necklace. "She kept glancing at you the whole way here."

"So?" Jim said as he started putting on his shirt.

"I'm just saying," Marie said. "What do you think that was about?"

"Hell if I know," Jim answered. "When I was single, there were several ladies that I was interested in. I always had a crush on Mary Tyler Moore, but she never expressed any interest in me.

"Why should I worry about someone who might, and I say might, be interested in me?" he replied. "There are probably several men who have been interested in you, but that doesn't matter.

"What matters is who I'm interested in," he continued as he pulled on his jeans. "And I'm only interested in one lady, you."

"And I'm only interested in one man," she said, kissing his cheek.

"Me?" Jim asked, pulling on his boots.

"No, silly," Marie answered. "Tom Selleck."

"I'm sure he's your second choice," Jim replied as he put on his sports coat.

"Of course," she replied. "Just like Mary Tyler Moore was your second choice."

"At least I can say that I beat out Tom for the prettiest lady I've ever met," Jim said as he looked at her. "Now, we need to head downstairs for the ride to the base. Maybe there'll be some others down there waiting, too."

As their driver navigated his way to Black Water headquarters, Marie asked, "Is this some secret location where there are security checks every ten feet?"

"Not really, but access is pretty strictly limited," Jim answered as they arrived at the entrance.

Once inside, the final guard instructed them to proceed to the dining area and find their table.

As Jim walked into the familiar room, he noticed it had been transformed into several round tables seating six people instead of the long tables he had seen every time before.

Finding their table, Jim spotted Gene at the front of the room, talking to one of the guys wearing the typical black knit shirt that designated Black Water members.

Walking over with Marie, he waited until Gene finished giving directions for the meal service.

"Marie," Gene said, turning to them. "How was the flight?"

"I think I can get used to that," she answered.

"Couldn't we all," he replied as he looked at all the people arriving. "I'm going to be rather tied up for a few minutes, making sure everything is ready before Mr. Hastings gets here. You guys wander around and just be ready to take your seats when I make the announcement."

Chapter 76

Jim led Marie back to their table just as Butch and Amanda found their table with them.

"Amanda, this is Marie," Jim said, introducing them.

"Marie, so nice to finally meet you," Amanda said, giving her a hug. "You look absolutely superb!"

"And so do you," Marie replied. "Again, thank you so much for helping me. I never imagined there would be so many people here."

"This was probably the biggest operation Black Water ever had," Butch said as they heard Gene asking everyone to please find their tables and be seated.

"Lots of people involved," Amanda said taking her seat beside Butch.

"Speaking of people, look who's here," Jim said as Minnie and Tammie walked over.

"Hi, Minnie," Jim said. "Please meet Marie.

"Marie, this is Minnie," he told her. "Minnie was my interpreter when I couldn't quite understand Mandarin."

"Not everyone only speaks some English, lots of Texan, and way too much profanity," Minnie said, shaking Marie's hand. "Nice to meet you."

"Please, let's take our seats," Gene announced from the podium again. "We need to get this started so Mr. Hastings can thank you for coming and get back to his busy schedule."

Minutes later, when everyone was in their seats, Gene said, "Ladies and gentlemen, please allow me to introduce this evening's speaker, the newly appointed Border Czar, and the man who most likely was behind the largest deportation operation this country has ever seen, Mr. Ted Hastings."

As Ted stepped to the microphone, he said, "Ladies and gentlemen, it's my great pleasure to have a chance to thank all of you, and those who couldn't be here with us this evening, for the outstanding work in helping rid this country of some of the most vicious criminals to ever come across our borders.

"The best estimates are that over four hundred thousand illegal immigrants with criminal histories have entered our country over the last four years due to the lack of border security," he told them.

"This operation, involving ICE, the FBI, Border Patrol, and other agencies, has resulted in the capture of over three hundred thousand of the most vicious of these people," he continued.

"I'm not going to single out any one person for the success of this operation," he said. "Everyone involved has my greatest respect for their service to this country.

"However, I will mention one group that provided the clue that led to the apprehension of several top members of the Chinese Triad," he continued.

"That group discovered an individual who claimed diplomatic status as a cultural attaché," he explained. "Once we learned that person's involvement with the massage parlors and the drug operation, we started looking for additional cultural attachés around the country.

"It's amazing how much culture we must have to warrant so many attachés from China," he said. "That led us to over one hundred of these attachés that we arrested.

"Now, even though they had diplomatic immunity, they were detained awaiting the Chinese to decide whether or not to waive immunity in accordance with the nineteen sixty-one Vienna Convention," he explained.

"We fully expect them to demand the release of these individuals, but in the meantime, we have a pocket full of aces," he continued.

"If you're familiar with the game of Texas Holdem, you know that pocket aces win about eighty-five percent of the time," he explained. "Well, we have basically an entire deck of aces.

"This will be used to get several of our people, and those of our closest allies, out of prisons around the world," he explained. "So, thanks to the enterprising people who made this discovery, our special thanks.

"And just as importantly, this will send a message to the rest of the world that the United States of America is closed unless people wanting to come here follow the same procedures we've used before this ill-conceived open border policy was allowed to replace legal immigration status," he told them.

"Now, if you'll excuse me, I've got the monumental task of figuring out how to get all of the almost one million illegal immigrants back to their home countries," he finished before nodding to Gene.

Gene walked over and shook his hand, saying, "Thank you for sparing the time to meet with these wonderful people, sir. It's been a pleasure having you here this evening."

"The pleasure is all mine," Ted said, shaking his hand. "This couldn't have happened if it weren't for these people."

Everyone stood and applauded as Ted left, waving as he was escorted from the room. As he left, they took their seats and waited for Gene to return to the podium.

"Okay," Gene said. "I'm going to apologize for both the crowded seating and the limited choice of meals for the evening, but with the short time we had to make arrangements for Mr. Hastings to come, this was the best we could do.

"So, please enjoy your meals, and there will be an open bar when you return to your hotel," he told them. "I'm sure I'll get a chance to come thank everyone of you when I'm done here and can make it over there."

"Who was the attaché he was talking about?" Marie asked as a team of waiters began delivering their meals.

"A man named Ming Li," Minnie told her. "And Jim found him."

Chapter 77

As most of the people had finished their meals, Gene stepped to the podium and announced, "Excuse me. But I have some important news to tell you."

Waiting for the murmur of the people to subside, he told them, "Mr. Hastings' convoy was attacked only minutes after leaving here.

"I'm happy to report that he is unharmed and has made it safely back to the White House," Gene explained. "Initial reports point to the Chinese Triad as being behind this attack.

"The group who attacked the convoy was using the QBZ-191, which is the latest fully automatic assault rifle the Chinese have developed," he continued. "Now, we suspect the attack was a response to our detaining a large number of their attachés, but we're taking an active part in making sure it doesn't include our people.

"So, as an act of extreme caution, we're sending teams to the hotel where you are staying and doing an extensive search of your rooms to make sure you are safe when you return," he told them. "Unfortunately, we've decided to cancel the open bar due to our inability to ensure our people aren't being targeted.

"So, while we are waiting for the teams to go through all of your rooms, we're bringing in a staff who will set up a small bar at the rear of the room," Gene continued. "It definitely won't be as well stocked as the one at the hotel, but since it could take a couple of hours to check all the rooms, it's the best we can do on short notice.

"I hope this inconvenience doesn't dampen your spirits for this celebration, but I encourage you to meet and talk to the folks from the different cities who performed the same outstanding work all of you did where you were assigned," Gene finished. "I'll be back to let you know when it's safe to head to the hotel. In the meantime, we're looking at alternate hotels and making arrangements for transportation."

"Do you really think it was the Triad?" Butch asked as Gene left the podium. "Couldn't it have been TDA?"

"I don't think the TDA is capable of the coordination to make an attack this quickly," Jim answered. "Plus, they are more of a smash-and-grab operation. This took some planning, and the weapons used aren't available here in the States in any great number.

"No, I think Gene is correct," Jim said. "This sounds more like the Triad."

"Do you think they know you were responsible for finding the first attaché and causing all of them to be found?" Marie asked. "And, what if they are targeting you now?"

"No, I don't think I'm a political target," Jim answered. "This game is being played well above the common foot soldier. I think this was more of a shot across the bow than anything else. They are playing chess, and this is their countermove to us taking their people.

"I'm sure they knew they couldn't kidnap Mr. Hastings and use him in trading for their people, but they couldn't just sit on their butts and do nothing," Jim said as he saw Debbie approaching their table.

"Debbie, how nice to see you again," Jim said, standing.

As he was giving her a quick hug, Jewell stepped from behind her, asking, "What, it's not nice to see me again?"

"Of course, Jewell," Jim told her, giving her a quick hug. "It's always nice to see you. By the way, my compliments to the tailor who sewed your head back together. Is that a cross stitch or a slip stitch?"

"Marie, I'd like to introduce you to a couple of special people whom I've had the pleasure of working with over so many years," Jim said as Marie stood beside him.

"First, this is Debbie," Jim said, introducing her. "Debbie, please say hello to Marie."

As the two women shook hands, Jim continued, "Debbie is the brains behind most of the high technology we've used in the field to find and identify the bad guys."

"Nice to meet you, Debbie," Marie said. "Are you the lady Jim said has developed some new facial recognition computer program?"

"Possibly," Debbie answered. "I'm not sure which program Jim was referring to, but I have developed a couple."

"I'll bet you're the one who provided the Dallas Police with the information that led to them finding the man who killed my husband," Marie told her.

"I did some work with them," Debbie acknowledged. "I've also worked with different police departments across the country, as well as with numerous federal agencies."

"Now, Marie, this is Jewell," Jim said. "She's the lady who was kidnapped and involved in the wreck on the video you've seen."

"Nice to meet you, Marie," Jewell said, shaking her hand.

"You, too," Marie said, looking at the stitches on her forehead. "I'm sorry about the accident, but I'm glad you're okay."

"You can blame the *accident* on Jim's driving," Jewell said, smiling.

"I wasn't driving," Jim responded, shaking his head. "But I'm beginning to think you're just unlucky with all of the *accidents* you've had over the years."

"Maybe being around you is what's unlucky," Jewell said, smiling. "Every time I've gotten hurt, it's been connected to some operation you've been associated with."

"All I can say is that I've been around Jim for quite some time, and I feel pretty lucky," Marie said, putting her arm around Jim's waist.

"Of course, you know I'm just giving him a hard time," Jewell replied. "Jim's always been a perfect gentleman and very professional. His sense of humor could use some tweaking, but I've always enjoyed working with him. All told, I'd say you are very lucky."

"Looks like they finally got the bar opened," Debbie said, looking at the rear of the room. "I'd like to buy all of you a drink, if you'll join me."

"That's so nice of you, Debbie," Marie said as they headed for the bar.

"Oh, Debbie's one of the most generous ladies with the company," Jim said, following Marie. "Especially when she knows it's the company's money she's generous with."

Later, after the all clear had been given for everyone to return to the hotel, Jim and Marie were lying in bed when she said, "I guess I owe you an apology."

"For what?" Jim asked as she put her arm across his chest.

"For thinking you were keeping something from me," she answered.

"I was," Jim told her. "But it's not because I don't want you to know. It's just that sometimes things can be said that jeopardize the mission. Especially when it involves as many people as this one."

"Well, I still want to apologize," Marie said, running her hand down Jim's stomach. "I should always trust you. And I want to add, I'm pretty damn proud of you."